DATE DUE

The Domestic Poultry Keeper

The Domestic Poultry Keeper

ERIC BAILEY

Photography by
Eric Soothill

BLANDFORD PRESS
POOLE · DORSET

First published in the U.K. 1985 by
Blandford Press, Link House, West Street,
Poole, Dorset BH15 1LL.

Copyright © 1984 Eric Bailey

Distributed in the United States by
Sterling Publishing Co., Inc.,
2 Park Avenue, New York, N.Y. 10016.

British Library Cataloguing in Publication Data

Bailey, Eric
   The domestic poultry keeper.
   1. Poultry—Amateurs' manuals
   I. Title
   636.5′083   SF487

ISBN 0 7137 1338 0

Typeset in 10/12pt Palladium by
Megaron Typesetting
Printed and bound in
Great Britain by
Butler & Tanner Ltd,
Frome and London

Contents

To *Thelma*

Preface

This book is not in any way intended to be a work of technical reference. It is produced solely as a work of general information to assist interested persons who are, or will be in the future, keeping hens in the back garden or on an allotment, whether it be a commercial strain of egg-layers and/or meat producers or, indeed, one of the beautiful pure and rare breeds of poultry that are so interesting.

All statistics mentioned in this book should be taken as a mean value and not as technical data, as specific calculations will vary according to the breed and size of fowl.

Eric Bailey
Sywell 1984

Acknowledgements

My grateful thanks are extended to the following people, who very kindly gave their permission for photographs to be taken of their rare and true breeds of poultry. These people are highly representative of today's dedicated 'fanciers', whose constant endeavour is to preserve and develop the rare and true breeds of poultry, some of which have histories dating back more than 2000 years.

Tom Bartlett, Sydney Brassington, Dr Clive Carefoot, Richard Carr, Mr R Carek, Chatsworth Estate Farmyard, Mr & Mrs H Critchlow, Arnold Fletcher, Keith Greenhow, Geoff Hamer, Ronnie Hart, Eric Harrison, Brian Hamsworth, Mr T Heginbotham, Donald Hoyle, Mr W Hoyle, Robert Hurst, Les Jones, D J Lowe & Sons, Dr Marland, Mr W Nicholls, Mr T W Oldcom, Mrs Pitman, Jan Rodrigues, Edmund Sandham, Frank Shepherd, Roy Sutcliffe, Mr P Wackett, Andrew Wetters.

I also wish to express my grateful thanks to my loving wife for her patience and understanding while I have been writing this book. Without her encouragement, it might never have been completed. I also wish to thank her for all the help given to me at various stages during the process of completing the manuscript, for reading, correcting and typing the copy and generally managing the poultry flock from time to time. Most of all, for helping me see it through to the end. Her help has been a major contribution to this work.

Introduction

The interest in the use of eggs over the centuries has gradually increased as their nutritional value has been more readily recognised. In the late nineteenth century, commercial poultry keepers in Britain became concerned that some £3 million was being spent annually by the Government on importing eggs for domestic consumption. Added to this concern was the effect that the purchase had on the balance of payments, especially with Europe, and the uproar created by the Opposition in Parliament. This led to the start of a new era of interest in, and prosperity for, the domestic poultry keeper.

With the numerous economic and domestic problems which exist today, many people, both young and old, are again considering the idea of becoming more self-sufficient, either by cultivating their back gardens or by taking on extra land in the form of an allotment. Many more people, however, are developing an interest in domestic poultry keeping and it is with these people in mind that this book has been written. The enjoyment of breeding hens and, for those with families, teaching children to share this knowledge and responsibility, can be a most rewarding pastime. At the time of writing this book, there are well over 1 million families in Britain alone who have become interested in domestic poultry keeping and each of these family units produces over 2000 eggs each year.

For hundreds of years, poultry have been kept by families all over the world and, nowadays, domestic poultry keeping is a worthwhile hobby, as well as a profitable business, as the increase in the number of domestic poultry breeders indicates. There is no better way to reap the rewards of your efforts than to sell your surplus eggs to friends and neighbours; they will not buy a fresher egg anywhere else. Apart from the satisfaction of producing your own food, as did most families during the war years, you are also helping to meet the nation's food demands.

The future of domestic poultry keeping is undoubtedly assured and it only remains for those who already have some poultry, and who have that extra piece of land, to increase their stocks, and for those who are taking it up for the first time to remember to combine proven laying strains with healthy stock, to ensure its continuance. In order to succeed, you must have a sound basic knowledge and recognise that good management will produce good results. Those who already keep domestic poultry will realise that selling all their surplus eggs can be profitable. The prices that home-produced fresh eggs are fetching nowadays can cover all your costs and even give you a profit. The increased demand for such eggs is surely because of their freshness and flavour, which means that people are prepared to pay that little extra. There is no better value and nothing more nourishing than a real home-produced fresh egg. Eggs contain carbohydrates, fats, proteins and all the necessary vitamins and minerals which are essential for strong healthy growth and freshness is a priceless quality. Furthermore, an egg is one of the cleanest foods available as it is protected from germs by its shell.

Apart from selling all the surplus eggs, there are additional profits to be made from domestic hens. For instance, after the hens have reached their first moult, which is usually when they are between 12 and 14 months old, they can be sold as boiling fowls. In addition, poultry manure is a valuable asset to the keen gardener and a profitable by-product of breeding poultry. The white part of the droppings consists of a nitrogenous material which is highly beneficial in growing brassicas, especially when mixed with straw and added to the soil on the vegetable patch. If the manure is to be stored, add it to the compost heap periodically, as this will provide the nitrogen essential for the decomposition of the compost. As a rough guide, it can be estimated that two dozen hens will produce approximately 1 ton of manure over a period of 12 months. This is indeed an additional benefit from breeding hens and one that costs nothing.

Those who wish to keep domestic fowls in the back garden or on an allotment should first of all contact the local environmental health officer to confirm that there are no bye-laws restricting the keeping of domestic poultry in the area. Also, if you live in council-owned property, check whether there are any local regulations to be observed.

This book will give you all the information you need in order to set up and maintain your stock of domestic poultry, whatever scale your operation may be, covering care and management in Part I, and describing all the different true breeds of fowl in Part II, to help you choose the best type to keep.

Chapter 1

Stock

Selection of stock will depend on your aims in poultry keeping. You may start with the idea of producing a few eggs for the family or you may be interested in breeding some of the rarer varieties of fowl. The scale of your operation will depend on the time and space available. Poultry enthusiasts generally fall into one of the following categories:

a) commercial poultry farmers who rear hens either solely for egg-production or for the table. At present, only a few people combine both.
b) people with sufficient space in the garden to keep a dozen or so fowls to provide the family with fresh eggs.
c) breeders who concentrate on the production of fertile eggs for rearing pullets. These are needed by the commercial and family egg-producer to maintain stock.
d) those people who have ambitions to win prizes for birds possessing outstanding qualities in both egg-production and breed characteristics.
e) the fanciers, who are exhibitors of both the popular and the more unusual, sometimes rare, breeds. They produce their fowls to conform as near as possible to the show standards of the Poultry Club.

There is ample scope for the domestic poultry keeper to breed poultry of all classes, but it is essential to obtain the correct breeding stock from reliable hatcheries which pay close attention to both the health and the laying qualities of their birds.

People who show an initial interest in keeping poultry invariably ask the question, 'Which are the best breeds of fowl to buy?'. There is no breed which possesses such outstanding qualities that it could be honestly described as the best breed obtainable. There are undoubtedly a good many breeds which are profitable, but the degree of profitability is determined by their breeding history. Those with little or no experience should be aware that the return from any breed of fowl depends on good husbandry and care taken in production and

11

development. Certain breeds are particularly noted for their qualities of egg-laying, meat-production or exhibition.

The most accurate method of testing the value of specific strains is to provide birds of each individual breed with equally favourable conditions and accommodation and to record their relative qualities over the years. Some domestic poultry keepers believe that they have a much better breed of fowl because they have acquired a flock of hens that has given better results than others which they have previously kept.

There are several factors which are important to success, one of which is the acclimatisation of the birds to the soil — it pays to rear the birds on the land which they are to occupy at maturity. (This is not always possible for domestic poultry keepers who start by buying breeding stock, because they usually purchase mature birds.)

Breed Comparisons

Chickens

It will be realised that no matter how good the strains, young chicks reared on the same tract of land as the breeding stock will invariably give better results than their parents, because they have become accustomed to the conditions while being reared.

It is interesting to note that the commercial breeds of today have evolved through cross-breeding of the true breeds of poultry for their utilitarian properties because, although more and more people like to keep fowls which are attractive in appearance, they are by no means prepared to keep them merely for their decorative properties. In the majority of cases, there are several varieties within the commercial breeds and the various qualities of these will vary in relation to the amount of care taken during their development.

Two of the most important points for success, both in rearing the chickens and in the number of eggs produced by the pullets, are stamina and fitness. For good results, the cocks should have been produced from good healthy stock and should show a marked virility and the hens or pullets should have been maintained in a controlled environment; in other words, good husbandry is essential at all times. The family history of the birds is important, whether the stock has been purchased for breeding, meat-production or egg-laying.

Those who are only interested in purchasing a few pullets with the object of producing eggs can usually obtain them from a commercial hatchery or pedigree breeder, as day-old chicks or at any age up to 18 weeks, i.e. at point-of-lay.

The person with a limited amount of land available must be certain to buy in the kind of birds most suited to confined quarters. For the newcomer to poultry keeping, perhaps the best way to start is to buy hens which are at point-of-lay. These are less troublesome and, within a week or so of feeding, you will already be getting results — i.e the first few eggs will already be rolling off your production line. If, however, you want your pullets to lay regularly as soon as you get them home, then you will have to pay more, as they will have to be about 6 months old. You can, of course, start with pullets of 10-12 weeks of age, which are much cheaper, but then you (rather than the breeder) will have to feed them to point-of-lay.

There are numerous suppliers of poultry of all ages (see Appendices). The most convenient method of finding your nearest poultry supplier is to look in the classified advertisements of your local newspaper under the heading of 'Livestock and Poultry'.

Unfortunately, the popularity of poultry keeping by the amateur today allows scope for the unprincipled breeder or dealer to offer for sale at private auctions or on obscure farms, pullets of indifferent quality. Stock purchased in this way can prove to be an unwise and unprofitable investment. The utmost care should therefore be taken when purchasing stock.

Many years ago, the three breeds which were recognised as the best laying birds were the Rhode Island Red, White Leghorn and the Sussex White. Nowadays, various strains of hybrid birds have been produced from these true breeds, purely for the purpose of egg-laying or meat-production.

On numerous occasions, amateurs have bought cross-bred pullets. The initial outlay is quite small, but by the time the birds have come into full lay, they find that, by calculating the cost of feeding and adding this to the initial purchase price, the birds have actually cost more than a specialised commercial breed would have done. It cannot be emphasised enough that it is not in the domestic poultry keeper's interest to purchase pullets which cannot produce the number of eggs necessary to show profitable results. The present-day commercial breeds can be divided into two types:

a) the general purpose table fowl,
b) those produced for egg-laying.

There are times when some will enjoy a greater popularity than others and the commercial breeds most commonly available in Britain are featured in Appendix 2. The qualities of the true breeds of poultry are shown in Table 1.

Table 1 True Breeds of Poultry

General Purpose Fowl

Barnevelder	Sussex
Orpington	Wyandotte
Rhode Island Red	

General Layers

Ancona	Minorca
Leghorn	Welsummer

Less Popular Breeds (mainly for exhibition)

Andalusian	Malay
Araucana	Maran
Appenzeller	Modern Game
Aseel (Asil)	New Hampshire Red
Brahma	Old English Game
Campine	Old English Pheasant Fowl
Cochin	Plymouth Rock (Barred)
Croad-Langshan	Poland
Dorking (Red)	Redcap
Faverolle	Red Jungle Fowl
Frizzle	Sicilian Buttercup
Hamburgh	Silkie
Houdan	Spanish
Jubilee Indian Game	Sumatra Game
La Fleche	Yokohama (Red Saddled)
Lakenfelder	

Many of the hybrid egg-layers are attractive and docile birds which are easily managed and will produce 4-5 eggs per bird per week on average. Most hybrids will perform well under all conditions, providing that the domestic poultry keeper adopts a good husbandry system. A small domestic stock can give just as good results as many of the best commercial units. A flock of hybrid egg-layers can quite confidently be expected to lay 280-300 eggs per bird in a 14-month laying cycle. Under good husbandry and depending on the time of year, most hybrids will lay for up to 14 months before the onset of the first moult.

It is useful to be able to compare the characteristics of different breeds when selecting stock and Table 2 gives a useful comparison of the number of eggs which average commercial layers will produce in their first laying cycle, together with the amount of food they will consume. From this information, the total feed bills can be calculated for a given period and related to the number of eggs which will be produced in that time.

Bantams

For those aspiring poultry keepers who have only a small back garden, it may be worth considering keeping bantams rather than the standard-sized domestic fowl. Bantams, being smaller, take up less space and also consume less food. They are almost certainly more economical to keep, but they lay smaller and fewer eggs and their carcass value is also correspondingly reduced.

Although they are almost certainly worth preparing for the table once they have ceased to lay, they are certainly not worth breeding for meat-production — one carcass will supply an average meal for one person! However, the eggs, although small, have a relatively large yolk and are usually fairly uniform in size. This makes them ideal for culinary purposes, especially if exact weights have to be adhered to. Prepared in the same way as standard eggs, they have a delightful flavour and texture. It should be possible to cover the cost of the bantams' keep by selling eggs, etc.

Development of the bantam fowl has made rapid progress over the last few decades and they are available in a greater range of colours and patterns than the standard domestic fowl.

Turkeys

Turkeys are temperamental, rather stupid at times and very demanding. Their constant gobbling sound can also prove an annoyance to neighbours. They need careful attention during the early stages of their development because they are more vulnerable to a variety of ailments. As they are generally reared solely with meat-production in mind, it is essential that they have a good balanced diet to sustain rapid growth. Bear these factors in mind and remember that the loss of one turkey out of six will almost certainly mean that it would have been cheaper to purchase your birds from a supermarket than rear them yourself.

Nevertheless, if you decide to go ahead and raise your own birds for the table, not only for Christmas but throughout the year, then the hybrid turkey poult will prove the most satisfactory for your purposes. As with chickens, hybrid strains are now produced in their

15

Table 2 Stock Comparisons

	Hybrids White Egg-Layers	*Specifications** Hybrids Tinted Egg-Layers*	*Hybrids Brown Egg-Layers*
Hen housed for an average of 72 weeks	260 eggs	270 eggs	245 eggs
Age at first egg	19-22 weeks	19-21 weeks	19-21 weeks
Age at peak laying	27-29 weeks	28-30 weeks	27-29 weeks
Large grade eggs over 63.7 gm (2¾ oz)	45%	41%	55%
Standard eggs over 53.2 gm (1⅞ oz)	45%	43%	37%
Shell colour	White	Tinted	Brown
Feed consumption 0-18 weeks	6.24 kg (13¾ lb)	6.8 kg (15 lb)	6.8 kg (15 lb)
Feed consumption 18-72 weeks	39.5-44.0 kg (87-97 lb)	43-46 kg (95-102 lb)	41.8-46.0 kg (92-102 lb)
Average feed per bird per day	106-118 gm (3¾-4¼ oz)	116-125 gm (4-4¼ oz)	111-122 gm (4-4½ oz)
Feed per dozen eggs	1.9 kg (4¼ lb)	2.0 kg (4 lb 6 oz)	2.0 kg (4 lb 6 oz)
Body weight at 18 weeks	1.2 kg (2½ lb)	1.5 kg (3¼ lb)	1.5 kg (3¼ lb)
Body weight at 72 weeks	1.9-2.0 kg (4 lb 3 oz-4 lb 6 oz)	2.15-2.35 kg (4¾ lb-5¼ lb)	1.95-2.13 kg) (4 lb 3 oz-4 lb 11 oz)

*All statistics are approximate.

thousands by commercial breeders and are readily available.

It is not advisable to purchase day-old chicks unless you have the use of a brooder, as they need a lot of attention and there is a high mortality risk. Hybrid poults of between 6 and 8 weeks of age no longer require artificial heat and they have passed through the high risk period.

If you intend to breed turkeys mainly for the Christmas period, the poults should be purchased in either June or July, depending on the weight required at the time of culling. The only drawback is that, during these 2 months, many of the smaller commercial breeders will be purchasing their stock and it is at this time that the price of a poult is at its highest.

If you do decide to buy day-old turkey poults from the breeder or hatchery, make sure that they will be dispatched in suitable containers (see Figure 1). These non-returnable boxes are constructed to a British Standards Institute specification. They are made of stout cardboard or corrugated fibreboard with reinforcing partitions. A circular strip of cardboard should be fitted into each compartment and suitable nesting material should be provided to give the young poults a feeling of security, as well as to separate them during transit. The boxes should be adequately ventilated at all times.

The boxes illustrated have small wooden battens fixed along the top of the lid, and the ends extend for 2.5 cm (1 in) beyond the sides; this guarantees that the boxes cannot be too closely packed and avoids any danger of the youngsters becoming overheated or suffocated during the journey.

There is no need for the young poults to be fed or watered before

Figure 1. Containers for transporting day-old turkey poults.

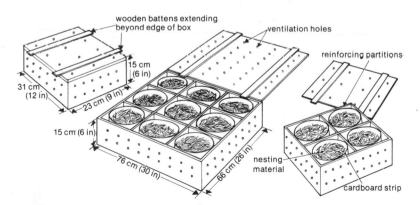

17

dispatch as there will be sufficient goodness in the yolk, which nature has placed within their digestive system, to sustain them for at least 36 hours. On no account should the journey take longer than 48 hours unless adequate arrangements have been made for the youngsters to be inspected after the second day.

The supplier should make sure that the poults have been properly packed and dispatch them immediately, after making sure that the box has been properly labelled with not only the name and address of the recipient, but also the telephone number and instructions as to how the box should be stored, i.e. away from draughts and fires.

Chapter 2

Accommodation

Housing

There are many good poultry houses on the market and, if you are planning to buy a ready-made house (Figure 2), do make sure that it is constructed of strong tongue-and-groove timber. It is not advisable to buy poultry houses that are made with the wooden slats similar to those used for fencing panels as they warp too easily.

Whether you propose to buy a house or construct your own, you should first decide how many hens you are planning to rear, as this will be the deciding factor in the size of poultry house required. As a

Figure 2. A typical poultry house showing nesting boxes with access from outside for easy egg collection. The door should be either solid or of the stable-type to allow easy access for cleaning purposes. A small aperture with door/ramp at the base of the house will allow the hens to come and go as they please.

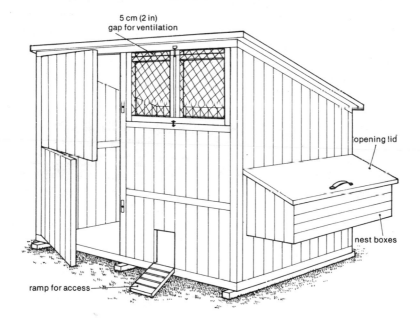

5 cm (2 in)
gap for ventilation

opening lid

nest boxes

ramp for access

19

rough guide, remember to allow an area of approximately 0.2 m²
(2 sq ft) per bird, bearing in mind that a little more space, up to
0.3 m² (3 sq ft), is needed for the larger breeds.

The most convenient size of poultry house has a floor area of
1.4×1.2 m (4½×4 ft) and a height of 1.4 m (4½ ft) at the front,
sloping to 1.2 m (4 ft) at the back. The house should have two floors:
the lower sliding floor should be solid and the upper sliding floor
slatted. The sides, both front and back, should be constructed of 18
mm (¾ in) tongue-and-groove matching board and the roof of 18
mm (¾ in) flooring board. The roof should be covered with a good
roofing felt and the inside and outside should be treated with
creosote. A hen house of this nature will accommodate up to 9 birds
and last for many, many years.

If the houses are not fixed on wheels, they should be raised up on
bricks, ensuring that there is good support underneath. By arranging
the house in this way, air will be able to circulate all round the
building and will assist in keeping both the house and the litter dry.

At the back of each house, there should be a hinged shutter or vent
which can be correctly adjusted, according to the weather conditions,
to maintain an even temperature inside the poultry house. Ventilation
is an important factor, as the removal of moisture from the building
is essential in poultry management. Chicken manure has a high
moisture content, sometimes as high as 75 per cent, and the hens'
respiratory moisture and the moisture in the incoming air must also
be taken into account. Good roof and wall insulation will help in
controlling excessive moisture. Contented hens, roosting in
comfortable warm houses, will repay their owners by producing
more eggs than those which are neglected and shivering in the cold
and draughts. It will be realised that, as climatic conditions can vary
considerably, and frequently, careful planning is essential.

Several poultry writers in the past, and unfortunately a few
amateurs today, still advance the theory that it is wiser to let birds
roost in open-fronted houses so that they do not notice the contrast in
temperature and thus benefit when let out in the morning. What they
do not realise, of course, is that the controlling factor in keeping out
the cold is the heat from the birds' bodies. Therefore, those fowls
which are supplied with plenty of protein and are situated in
comfortable housing conditions are the ones that will most enjoy their
freedom when let out in the mornings. The fowls which have roosted
in cold damp conditions will not enjoy their liberty when they get it.
The natural heat built up and maintained by the birds themselves,
when given the right type of feed and good housing, is a prerequisite

to the production of good healthy fowls which will enjoy being out of doors. If the hens are happy, they will continue to give their maximum egg-lay for a period of about 14 months but if their accommodation is open to the elements, they will be miserable and loss of profit will result.

Experience has been the guide to success in many an occupation and nowhere is it more valuable than poultry keeping on a small scale. Experiments have shown that open-fronted buildings are just not suitable. In the colder months, it is necessary to enclose the fronts of the houses not only to provide additional warmth for the hens but to maintain egg-production. It has been proved that, when subjected to a misty cold atmosphere, hens require all their food to maintain body temperature, leaving no surplus for egg-production until warmer weather conditions prevail.

If you are planning to buy a new hen house, then the initial cost can be rather high due to the price of timber. If you are a keen do-it-yourselfer or joiner, however, you can reduce your costs considerably by building your own. If you are lucky, you may be able to purchase a second-hand unit, but these are very seldom available. The place to look is in the 'Articles For Sale' column of the local paper.

Housing For Breeding Stock and Layers

Both breeding stock and domestic laying fowls need similar accommodation — dry and warm with plenty of air space above them so as to avoid an unpleasant atmosphere. It is a wise poultry keeper who avoids the mistake of having cold housing conditions which can have an undermining effect on the birds' constitutions.

Hens will go to roost as soon as it is dusk and, just as they do after feeding, will generate some heat when on the perch. The house, if constructed of good substantial timber, and with an opening of 7.5-10 cm (3-4 in) under the roof at the front to provide a reasonable amount of ventilation, will produce no unpleasant odours, providing that it remains dry. This form of ventilation is all that is required during the autumn and winter months. Ideally, there should also be windows, which can be opened during the warmer periods of the year, thus giving the poultry keeper year-round control over the ventilation.

Positioning

The position of the hen house in relation to dwelling places is most important as there is the possibility that, from time to time, the smell

21

from the poultry house and run might be noticeable. The house should therefore be situated so that the prevailing winds do not carry any odours towards your home, or to the homes of your neighbours. A position on a slight slope facing south is ideal as this allows maximum light into the house and also allows for good drainage. Although hens are not very susceptible to the cold, they do dislike mud and wind. They will continue to lay right through the coldest weather but, if conditions are damp and windy, then egg production will suffer. Suitable housing can minimise this problem.

Fold Units

An alternative form of housing hens is the fold unit (Figure 3). These are compact portable hen houses that are useful when space is limited. It is, however, essential that these fold units are moved to fresh ground every 2 to 3 days, even daily during inclement weather. This task can be carried out at feeding time each morning.

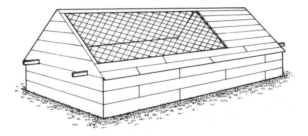

Figure 3. A movable fold unit designed for easy handling so that the hens can be moved to fresh ground each day.

Nesting Boxes

Laying stock will, of course, require nesting boxes in which to lay their eggs. If a ready-made poultry house is purchased from the manufacturer, then, invariably, there will be nesting boxes built in, forming part of the building. If, however, this is not the case then it will be necessary for the domestic poultry keeper to construct them himself. This is not a difficult task and reference to Figure 4 will

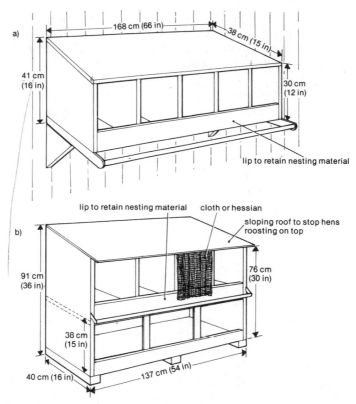

Figure 4. Nesting boxes suitable for a large hen house. These can be: (a) fixed to the wall or (b) free standing. Both have a wooden lip at the front to avoid the nesting material being pushed out. The top tier in (b) consists of separate sections whereas the bottom tier is communal.

indicate how best to tackle the job. If nesting boxes are not made available to the laying stock, they will simply lay their eggs on the floor of the house or any other place that they find suitable. This is not very satisfactory as the eggs may become contaminated with bacteria. It also makes collecting the eggs more difficult.

There are two types of nesting boxes: individual and communal. If individual boxes are opted for, one nesting box should be allowed for every three hens; the area of each nest should be about 0.1 m² (1 sq ft), not less, and the box should be lined with a good supply of nesting material, e.g. wheat straw or wood shavings to enable the bird to make its own nest. If communal nesting boxes are used, they should be of sufficient length to support six or eight hens. These, however, are not quite as satisfactory as individual nesting boxes, although they have the advantage of being easier to construct. Whichever type of nesting material is used, do make sure that the nests are kept clean by inspecting their condition at each egg collection; if they are allowed to become dirty then so will the eggs laid in them.

Hens do like solitude when laying their eggs and therefore it is important that nesting boxes be reasonably dark inside. They should also be constructed and positioned in such a way that the hens using them will still have a view of the rest of the flock — they are curious creatures and like to see what is going on around them. The nests can be darkened by hanging strips of cloth or sacking from the front eaves (Figure 4).

Perches

When dusk begins to fall, it is the natural instinct of hens to come home to roost. Therefore, although it is not absolutely essential, a wise poultry keeper will supply his flock with perches on which to roost during the hours of darkness. Allowing the hens access to perches has several advantages: it gives them a greater sense of security and they are therefore more comfortable; it avoids all of the hens crowding together in one corner of the hen house, which cuts down the circulation of air around their bodies, rendering the birds more susceptible to infection or other ailments. Perches have the effect of keeping the birds together at night but still allowing the air to circulate around them.

If perches are made available, do make sure that they are positioned correctly, for there is a right and a wrong way (Figure 5).

Figure 5. Hens do prefer to roost close together at night, therefore perches should be positioned not in steps as illustrated in (a) but all at the same level, as in (b).

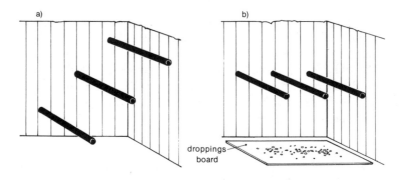

The perches can be constructed of 38×50 mm (1½×2 in) timber, with the two upper edges being rounded off to make it easier and more comfortable for the hens to grip with their feet. They should all be positioned at the same level, approximately 40 cm (16 in) apart. Ensure that there is sufficient perching space for the entire flock by allowing 15-20 cm (6-8 in) of space per bird.

Hens produce about 50 per cent of their droppings whilst they are roosting on the perch. It is therefore advantageous to have a droppings board positioned immediately beneath the perches and overlapping the area by about 23 cm (9 in) or so. This simplifies the job of cleaning the house, as the droppings can be scraped off the board once a week and added to the compost heap. The use of a droppings board allows the rest of the floor area to be left clear, so that the birds can exercise in the straw litter on the floor of the house, especially during inclement weather. The excreta produced by hens is rich in nitrogen and a valuable asset to the keen gardener, as it is an ideal fertiliser for growing all brassicas and leguminous plants.

Runs

Hens do love their freedom and the ideal way in which they can enjoy this luxury is to allow them free-range, i.e. to give them *carte blanche* to wander wherever they please (except on the cabbage patch for they will devour everything in sight!). However, it is not always possible for the domestic poultry keeper to find that extra 1 or 2 acres of land on which to allow them to roam freely. It is therefore necessary to consider how best to house the flock in a small area.

The dimensions of poultry runs can be varied according to the number of fowls kept and the amount of space available. As a rule-of-thumb, allow at the very least 0.3 m² (3 sq ft) of run area per bird, plus the area of the hen house. For example, a flock of twelve hens housed in a building 2.4×1.8 m (8×6 ft) will require a run at least 3.7×2 m (12×7 ft) in area. If the run can be extended to twice the size, it can be divided into two sections (Figure 6) with the use of a portable dividing fence; this is a great help if the run is to be constructed on a grass area, which will soon become a quagmire in inclement weather. By dividing the pen, it will then become possible to rest one section every few weeks.

The simplest way of constructing the run is to use 38 or 50 mm (1 or 2 in) wire mesh (chicken wire) and pine posts 2.4 m (8 ft) long. Mark out the dimensions of the run and then drive the posts (the ends of which should have been tapered and creosoted) firmly into the

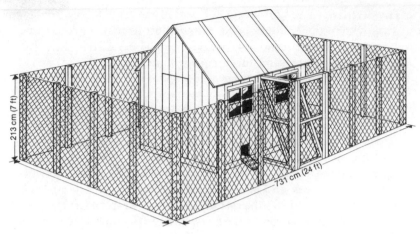

Figure 6. A useful poultry house and run suitable for an average-sized rural back-garden. The run can be divided into two sections so that the hens can run in one side while the ground in the other recovers.

ground to a depth of 30-45 cm (1-1½ ft) at intervals of 2.1-2.4 m (7-8 ft). The wire mesh can then be stapled onto the posts or fixed with metal ties so that it can be easily removed at any time. The bottom 15-30 cm (6-12 in) of the chicken wire should be buried in the ground, extending outwards from the run; this will stop the sly old fox or other animals from digging their way underneath, which happens, especially during the winter months. When fixing the chicken wire, do make sure that it is kept taut and remember to allow for a small wooden gate to be incorporated into the fencing for access. The corner posts will require supporting stakes to that they are not pulled out of true by the tightness of the wire mesh.

Hens kept in a confined area will soon denude the ground of greenery. Therefore it is advisable, although not absolutely essential if the birds are on a balanced diet and *ad lib* feeding, to supply them periodically with weeds from the garden, spent Brussels sprout stalks and the like. This helps to keep them occupied and prevents the habit of feather-eating developing. If possible, fix the Brussels sprout stalks so that they hang just above the hens' heads; this makes them jump up thus giving them plenty of exercise and keeping them busy.

Housing for Bantams

The cost of housing, like that of other aspects of poultry keeping, has increased over the past years and savings can be made if the poultry keeper decides to keep bantams. Being smaller in size and lighter in weight, their housing, nesting boxes and runs can be correspondingly smaller. Generally speaking, the nesting and sleeping accommodation required for, say, twelve bantam hens need be no larger than 1.8 × 1.2 m (6 × 4 ft) with a height of just 90 cm (3 ft) at the front and only 30 cm (1 ft) at the back. The run, which is attached to the house, need be no more than 3.7 × 1.8 m (12 × 6 ft) in area with a fence 60 cm (2 ft high).

Housing for Turkeys

Poults that are 6 weeks old and upward can be housed in much the same way as chickens, the main difference being that the entrances and exits must be proportionately larger to allow for their greater size. Turkeys will also make good use of perches, providing that they are not too high, but it is not essential to provide them with such luxuries. Neither is it necessary to provide them with nest boxes, unless it is intended to rear poults — an operation which can be rather costly. When deciding where the turkeys are to be kept, do remember that each bird requires an average of 0.37 m² (4 sq ft) of floor space and double that area for the outside run. Turkeys can, of course, be kept together with domestic hens in the same accommodation, providing both are housed at the same time and at the same age, otherwise fighting will occur. However, separation of the flocks is always advisable, simply because of the differences in the balanced diet once the chickens have reached point-of-lay.

Turkeys are rather stupid at times and, if they are outside of their house when darkness falls, they will invariably just squat where they stand and remain there until daylight. Should it start to rain, they will not even have the intelligence to run indoors and protect themselves from the wet. (If they are allowed to remain outside in wet conditions they will soon become chilled, which can lead to respiratory ailments.) Because of their stupid natures, it is always best to confine them in a dry, draught-proof building, which should also be frost-proof and rat free. Nevertheless, if the weather is fine, warm and dry, there is no reason at all why they should not be allowed out into the fresh air to enjoy the sunshine.

Chapter 3

Feeding

The digestive system of the hen is built on quite different lines (Figure 7) to that of mammals, of which Man is one. Most obviously, there are no teeth with which to chew the food and this passes directly into the oesophagus on swallowing. The oesophagus has become dilated to

Figure 7. The hen's body is a most intricate and highly efficient mechanism for egg production.

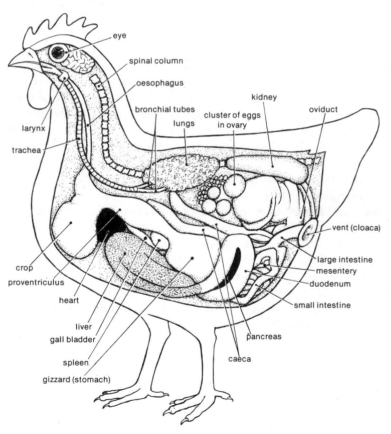

eye

spinal column

oesophagus

kidney

bronchial tubes

cluster of eggs in ovary

oviduct

lungs

larynx

trachea

vent (cloaca)

crop

large intestine

mesentery

proventriculus

duodenum

heart

small intestine

liver

gall bladder

pancreas

spleen

caeca

gizzard (stomach)

form the crop which acts as a storage organ. From the crop, food passes into the stomach: firstly into the proventriculus, which is a glandular area secreting enzymes to break down the food, and secondly into the highly muscular gizzard, which grinds up the food, aided by grit and small pebbles swallowed by the bird. The small particles of food then pass into the intestines. In the first part of the intestines, secretions from the liver and the pancreas break down the food particles still more and, further along, the nutrients are absorbed into the bloodstream through the intestinal wall. Waste material passes into the rectum where water is re-absorbed and, finally, the solid waste, mixed with uric acid (the waste product from the kidneys — birds do not produce liquid urine) are expelled from the cloaca as the characteristic droppings.

Feeding Standard Poultry

Many different methods of feeding are advocated and the domestic poultry keeper may sometimes be puzzled as to which will produce the best results. Bear in mind that, unless the birds are given the correct food, the production and development of the eggs will be restricted. Correct feeding is much more important than is generally recognised by the amateur. Some breeders provide their hens with wet mash; others supply dry mash. The latter is a combination of several finely-ground cereals supplemented with either fish or meat meal — this is given to the hens in a dry state, either in hoppers or troughs (Figure 8).

It should be remembered that the well-being of the hens has to be provided for before there is any surplus of nutrients for egg-production. The correct food is essential and as the nourishment which it contains is only available to the hen after the food has been completely assimilated, some thought should be given to the fact that fowls should have foods which can be easily digested.

As a rough guide, most commercial hybrid strains will eat about 3 kg (7 lb) of layers' mash every 4 weeks. It is, of course, essential that the birds are allowed to eat as much as they need of a balanced diet and the best way to achieve this is to adopt the system of *ad lib* feeding. The actual amount of food eaten will vary according to the weather conditions, the rate of egg-lay and also the age of the birds. If the average winter temperatures are approximately 7°C (20°F) below the summer temperatures, then feed consumption in the winter could increase by about 14-22 gm (½ oz-¾ oz) per bird per day. Food wastage can also account for about 5 to 10 per cent of all feed used. If,

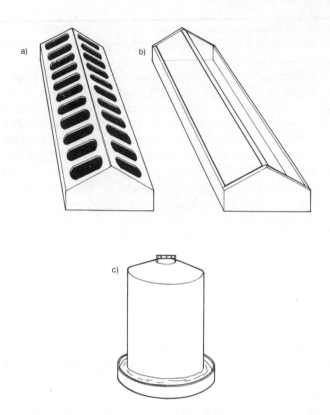

Figure 8. Essential equipment for everyday use: (a) chick feeder (b) adult feeder (c) automatic drinker — the central reservoir is filled with water and automatically replenishes the drinking area once the level falls below the air inlet or valve.

however, one uses a suitable type of feeding trough or hopper, then this wastage can be considerably reduced.

Do not make the mistake of thinking that a few hens in your back garden will give you an unlimited supply of eggs in return for the scraps of food left from your table, as such a diet would be deficient in many essential ingredients. Scraps may be used as part of a balanced diet, but remember that birds cannot tolerate excess salt and select the scraps accordingly.

Barley, which is usually the cheapest of grains, or a combination of mixed corn, normally wheat or kibbled maize, can be fed to the birds for their evening meal whatever the food programme you have decided to use and especially if the birds are on a dry mash diet. If grain is fed, not more than 14 gm (½ oz) per bird per day should be given as an excess can upset the balance of a regular layer's diet.

Hens need a balanced diet containing not only fats, proteins and carbohydrates but also vitamins and essential minerals. Such a diet is normally made up of grains which are ground and then blended with oils and minerals and supplemented with vitamins and other essential ingredients.

In this modern age, a balanced feed can be readily purchased from the local corn merchant or directly from the manufacturer's mill, depending on the quantity required. Most poultry feed, e.g. an intensive layers' mash, standard layers' mash or growers' mash, is packed in 25 kg (55 lb) bags. This balanced feed is also available in pellet form, which is much easier to use. It does cost a fraction more than the mash, but it works out cheaper in the long run as there is less wastage. On the other hand, you may prefer to blend your own poultry feed. Whichever type of feed is used, it is first necessary to decide on a daily feeding programme. This must be strictly adhered to, as variations in feed can upset the balance of the layers' diet, resulting in a drop in egg-production.

A well-balanced poultry feed invariably consists of most, if not all, of the following ingredients, in varying proportions depending on the age and type of fowl:

Wheat and/or Wheat Fibre
Barley
Maize (kibbled or crushed)
Soya and/or soya extract with full fat content
Sunflower
Limestone (powdered)
Meat and bone (powdered)
Mineral salts
Amino acids
Fish meal (sometimes)
Di-calcium phosphates
Colouring pigment (to give a darker yolk colour)
Trace elements, e.g. iron, phosphorus, manganese.

In addition to the above, there will be the necessary vitamins A, B_1 and B_2, D, E & K, and minerals, which are vital for strong healthy growth. These vitamins will make up about 1% of the general diet.

There are two methods used in manufacturing a balanced poultry food. Either all or most of the ingredients are mixed together in their correct proportions and then ground in the mill or each ingredient is ground separately in quantity and then blended in the correct proportions, with the vitamins and minerals being added separately afterwards. Whichever feeding method is adopted, apart from *ad lib* feeding, the following programme will generally produce good results.

Morning Feed (7 am-8am or one hour after sunrise)
Dry mash Of the appropriate balanced mash, feed approximately 85-113 gm (3-4 oz) per bird per day. Place the feed in a trough and make sure that each bird is able to have its share.
OR
A full feed of food in pellet form.
Wet mash Take 56-85 gm (2-3 oz) of the balanced mash per bird, place in a suitable container and add a sufficient quantity of clean water to moisten. Mix with a suitable utensil until it is slightly binding in the hand. Leave for 15 minutes or so and then feed to the hens, placing it in a suitable feeder or trough.

Afternoon Feed (3pm-4pm or at least one hour before sunset)
Feed as indicated above but to this can be added some of the household scraps. Making sure that it is well boiled beforehand and then mixed with the meal. The wet mash system is an ideal way of using household scraps as it blends quite easily. It must, however, be made clear that successful results depend on a properly balanced diet.
OR
A full feed of balanced pelleted food.
OR
A handful of grain per bird, either wheat or barley — barley being the cheaper of the two grains. This can be scattered over a grassy or clean area, which will make sure the birds will scratch about, thus getting plenty of exercise.

If the wet mash system is adopted it is important to make sure that the receptacles are thoroughly cleaned each day, to avoid the possibility of staleness or mould developing, this may well cause health problems.

If *ad lib* feeding is adopted, i.e. allowing the poultry access to the feed 24 hours per day, then general periodical topping-up is all that is required.

Surprisingly, the birds will never overeat as they will consume only sufficient food to meet their normal body requirements and allow for the production of eggs. For this method of feeding, the correct balanced mash of pellets should be purchased from a manufacturer or corn merchant. You should bear in mind that feed left lying about is likely to attract rats — a point to consider if you decide to adopt this *ad lib* feeding method.

Although it is admissible to allow the flock access to greenstuffs, in order to add variety to their diet, it is not essential: by using the properly balanced mash or pellets they will be receiving all the necessary vitamins and minerals that they need. If the birds are in a confined area, however, then they are in the ideal situation to receive all the spare cabbage leaves, lettuce, spent Brussels sprout stalks and all the weeds from the garden — an ideal way of giving them occupational therapy. Remember an active fowl is a healthy fowl.

It is possible, in some areas, that supply of manufactured and properly balanced poultry feed might not be readily available. The following guide to the proportions of ingredients may, therefore, be of some help when mixing a simplified balanced feed:

Bran 8 kg (18 lb)
Flaked or crushed maize 8 kg (18 lb)
Crushed wheat (with fibre) or crushed soya beans 5.5 kg (12 lb)
White fishmeal 2 kg (4 lb) — not more
Ground limestone 1.5 kg (3 lb)
TOTAL: 25 kg (55 lb)

If the above mixture is used then access to greenstuffs must be readily available. Preferably, the hens should be on free range, as in this way they can obtain their vitamins and minerals from natural sources.

Feeding Chicks
Baby chicks, once they have absorbed the yolk, which normally takes between 36 and 48 hours after hatching, should be allowed free access to the appropriate food at all times; this will maintain a steady growth rate. The following may be of some help to those who propose to rear their own chicks from time to time.

a) Chicks aged 2 days to 6 weeks should have a constant supply of balanced chick crumbs.
b) From the age of 6 weeks to point-of-lay, i.e. 18 weeks old, they should be fed standard rearers' or growers' mash or pellets.
c) At 18 weeks old, their diet should be changed for the last time to a standard or intensive layers' mash or pellets.

Grit

It is necessary to make freely available a supply of grit as this helps the hens to digest their food. The grit, which remains in the gizzard for some time, serves the function of teeth in birds. With the muscular movement of the gizzard, the grit acts as a grinding surface and reduces any large pieces of food down to smaller particles. The grain fed in the afternoon to mature hens can be ground by the normal muscular action alone, but the grit greatly increases the area of the grinding surface. Once the food has been ground down to tiny particles, it can then be effectively treated by the natural digestive juices of the hen.

Supplying the hens with grit also gives them a ready supply of calcium which is another essential feature of the diet. Ground oyster shell and flint dust are excellent minerals for laying hens as their use is two-fold. Firstly, they supply the material for making the shell (calcium) and, secondly, they help to digest the food. A little bone meal can also be given as this can assist in the development of the egg-producing organs as well as helping the body generally. Flint dust mixed with the mash is always an invaluable additive for laying hens, as it helps to prevent them from laying shell-less eggs. Flint dust is particularly useful if you intend to use some of the eggs for breeding (providing that they are fertilised of course) because it makes the shells of the eggs more brittle and so the young chicks are able to release themselves from the shell more easily.

Feeding Bantams

On average, the normal feed consumption of a bantam is about half that of a commercial hybrid hen and a bantam hen will lay approximately 12 dozen eggs during a complete laying cycle compared with 21 dozen from a standard hybrid layer. The feeding programme is otherwise much the same as for any other breed of standard fowl. Bantams will also eat exactly the same delicacies, i.e. pasture grass, spent cabbages and the like, tomatoes and any other tasty item that may come their way.

Feeding Turkeys

First and foremost, it must be decided whether the turkeys are to be reared for the table, i.e. for meat-production, or for egg-laying. Whatever the decision, for the first 20 weeks, the feeding programme for all poultry is quite a simple operation. Nevertheless, it does require some thought as all food supplied represents a capital expenditure. Therefore, to achieve the best possible results, the flock must be fed a properly balanced diet that has been blended for the particular purpose. Just as the food for standard laying hens must contain all the necessary vitamins and minerals, so too must the turkey rations; vitamins B1 (thiamin) and B2 (riboflavin), and minerals, including manganese, together with protein, in the correct proportions must all be included if success is to be achieved.

If you decide to purchase commercially blended feed from the manufacturer or the local corn merchant, remember that, if feed is purchased in bulk and stored for any length of time, the vitamins in the feed will lose some of their nutritional value after 2 or 3 months and the feed generally will become rancid and unappetizing. You should therefore regulate the quantities you buy so that the food will always be fresh and will give maximum benefit. If they are fed correctly, rearing domestic turkeys can be a reasonably profitable venture. The succulent white meat will be full flavoured and far more appetizing than the flesh of those birds which are mass-produced solely for the freezer.

If the birds are to be reared purely for meat-production, then an *ad lib* feeding programme is the most satisfactory. If, on the other hand, they are to be reared for breeding purposes, then regulated feeding is advantageous as the danger of over-feeding can be avoided — it is essential that turkey breeding stock should be brought into lay free from excess fat. The correct feed formulation will help to produce uniform growth and healthy stock suited for maximum meat-production or for egg-laying, with good fertility and hatchability during the breeding season.

Most breeding stock requires only two meals a day; to overfeed only produces excess fat. If the turkeys are kept in a confined area, then a dry or wet mash, or pelleted feed, should be given to them early in the morning (it is not recommended that they be allowed free-range); they will not require their next feed until an hour before sunset when they should be given a good handful or two of grain. The quantity will vary according to the bird, as some will always eat more than others, depending on the time of year and whether they are in lay. A good poultry keeper will soon be able to judge the correct quantity by trial and error. The simplest rule is to give the birds just as much as they can eat in 5 minutes, which is ample time for them to fill their crops. Most poultry will eat only sufficient for their needs and, if a surplus of grain is given, which happens especially when grain is fed outside the house, the remainder will just be left. It will then be eaten by the delighted sparrows and buntings — grateful for a free meal at your expense — as well as attracting vermin.

Choosing a Feeding Programme
Turkeys can be fed in much the same way as the general domestic hen, up to and including 20 weeks of age. Assuming that the turkeys are being bred for meat-production only, the next decision is at what age and at what weight are the birds required to be killed. With good husbandry and correct feeding a turkey at say 12 weeks of age will most probably weigh in the region of 4.5 kg (10 lb); at 16 weeks of age, it should be 6.4-7.3 kg (14-16 lb) in weight and at 24 weeks and upwards, it should weigh about 9 kg (20 lb) or more. These weights of course will depend on the breed and the type of food that has been given. Table 3 should assist the breeder in determining the most suitable feeding programme for his requirements.

Table 3 Feeding Programme for Meat-Production in Turkeys				
Feed		Weeks of Age		
Turkey starter crumbs	0-5	0-5	0-4	0-4
Turkey rearers' pellets	5-9	5-9	4-8	4-10
Turkey growers' and finisher pellets	9	9-14	8-13	10-17
Turkey topweight pellets		14	13	17
Age at Culling	10-15	16-20	21-24	21-24

On average one 25 kg (55 lb) bag of turkey growers' pellets will provide enough food for six birds of approximately 16 weeks of age for 12 to 14 days, depending on the time of year, as they always consume more feed during the colder months.

Turkeys that are to be reared as breeding stock will obviously require a different blend of feed to that which is used for meat-producing purposes. This is known as breeders' mash or breeders' pellets; the main difference is the lower proportion of protein in the feed. It is a commercially-blended poultry feed formulated to produce uniform growth and healthy birds capable of maximum egg-production with good fertility and hatchability in the breeding season. It will invariably also include anti-coccidiosis and anti-blackhead supplements.

Table 4 Feeding Programme for Breeding Turkeys *(ad lib)*

Feed	Weeks of Age	
	Stags	*Hens*
Turkey starter crumbs	0-6	0-6
Turkey rearers' mash/pellets	6-11	6-16
Turkey growers' mash/pellets	11-28	11-19 (point-of-lay)
Turkey breeders' mash/pellets	28	19-28
		(point-of-maximum-lay)
Turkey breeders' mash/pellets		28 —

The breeder can of course make up his own blend of turkey rations by purchasing the ingredients individually from the millers and mixing them in the correct proportions. This, however, is a rather laborious task as well as being time-consuming and is not necessarily any cheaper; on the contrary, it can prove more expensive than buying in the ready-mixed feeds produced especially for the purpose.

Whatever the type of feeding programme chosen, do remember that fresh clean drinking water must be available to the flock 24 hours a day (Figure 8). Flint and granite grit should also be made available, either sprinkled on the feed or supplied in a separate container next to the feeding trough or hopper. Oyster shell and limestone grit should never be fed to birds when a balanced or commercially-blended feed is being given as the requisite amount of calcium is already included in the ration.

Chapter 4

Eggs and Egg Production

What is an egg?

An egg is an item of food which the majority of people eat and enjoy without giving another thought as to how it is made — and yet it is one of the most sophisticated products of the natural world (Figure 9).

The yolk of an egg provides all the nutritional requirements of the young chick for the first 36 to 48 hours of its life and contains a range of nutrients, vitamins and minerals which make it a very rich and valuable food source for other animals. An average-sized fresh egg has an energy value of 80-85 calories and contains 12.5 per cent protein, 10 per cent fat and 65 per cent water; the remaining 12.5 per cent consists of minerals such as calcium for healthy bones, trace elements, such as magnesium, potassium and iron, and Vitamins A, B_1 and B_2, D, E and K.

In the wild, but only at certain times of the year, eggs are laid by all types of birds solely for the purpose of reproduction. In this modern age of commercial hybrid poultry farming, however, hens have been produced which will lay continuously day after day for a period of almost 14 months.

Figure 9. Structure of a fertilised egg. One of the most intricate mechanisms of evolution that nature has ever produced, a fertilised egg contains all the genetic material necessary to ensure the survival of the species.

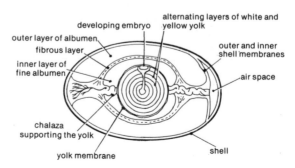

Because of the high water content of the egg, it is essential that a supply of clean drinking water is available at all times to the hens. They usually prefer water with the chill off it. In frosty weather, warm water should be given so that it does not ice over.

How An Egg is Made

To understand how an egg is produced, it is necessary to have some idea of the structure and workings of the hen's reproductive system. The position of these organs in relation to the rest of the body is shown in Figure 7. The production of egg cells is stimulated initially by hormones carried in the bloodstream. The egg cells, which are extremely yolky, are produced in one of the paired ovaries, thin sac-like membranes situated high up in the body near the centre of the back. As the egg cells develop and grow, they hang in a cluster from one ovary, almost like a bunch of grapes, but with the smaller ones (not larger than a pin head) at the top and the larger ones at the bottom. The largest egg cell, when it is fully developed, or ripened, detaches itself from the ovary and drops into a funnel at the top of the oviduct or egg passage. (If it encounters a male sperm at this stage, it will be fertilised and will eventually develop into a chick.)

The oviduct is a coiled tube about 40 cm (16 in) long, according to the size of the hen. During the first part of its journey down the oviduct, the yolky egg cell turns continuously and gathers albumen (egg white) from glands in the oviduct wall. If you could examine the yolk in the ovary, you would notice a few small blood vessels. These normally disappear before the egg passes into the oviduct but, if for any reason — such as the hen being startled — they rupture, tiny dark spots may appear in the white of the egg. During the next stage of the journey, the egg acquires protective membranes and, finally, a hard calcareous shell is secreted around it. These stages are shown in Figure 10. Its passage through the vent into the nest is assisted by mucous secretions of the oviduct. It normally takes between 18 and 20 hours for an egg to leave the ovary and be deposited in the nest although there are the occasional exceptions.

The Use of Artificial Light

To encourage the production of eggs during the shorter days of winter, an electric light can be installed in the hen house. One 25 watt bulb is sufficient to give enough light for a poultry house approximately 1.5×1.2 m (5×4 ft). For hens to maintain their

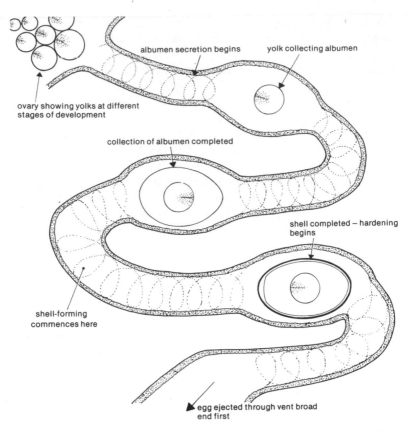

albumen secretion begins

yolk collecting albumen

ovary showing yolks at different stages of development

collection of albumen completed

shell completed – hardening begins

shell-forming commences here

egg ejected through vent broad end first

Figure 10. The development of an egg. The yolk falls from the ovary into the oviduct and collects albumen and a shell on its spiral journey to the vent.

maximum lay, a period of 15 hours light must be maintained; below this, egg-production will be reduced. From April to the middle of June, natural day length gradually increases from 14 hours to 17 or 18 hours, and after this, gradually decreases to approximately 8 hours in the shortest days of winter, before gradually increasing again. The poultry keepers' aim is to ensure that the birds are exposed to the maximum amount of light by supplementing natural daylight with artificial light when necessary. Birds which are 6 months old, i.e. at point-of-lay, when the hours of natural daylight are below 15, will give the maximum egg-production for the poultry keeper. When the birds are in lay, the hours of light they receive each day should by no means be decreased. Inexpensive time clocks can be fitted which can be adjusted to ensure that the correct amount of light is given. Most

40

hatcheries would be pleased to supply a lighting programme to facilitate the maximum production of eggs.

If no artificial lighting is provided, the best time to buy-in point-of-lay pullets would be in Autumn (October and November in Britain). Pullets purchased at this time of year will not suffer from the effects of diminishing light since the days gradually lengthen from 21 December. Re-stocking with day-old chicks in May should give similar results.

Broody Hens and Their Treatment

Although not many hens within the numerous strains of commercial breeds of poultry today develop broodiness, it does sometimes occur — as it did with one batch of hens being bred by the Author. As it may happen in general domestic poultry, and because egg-production may suffer as a result, the following information and suggestions for remedying the situation are offered. Of course, if you are intending to use the broody hen for rearing chicks, treatment will not be necessary.

When a particular hen is found to be sitting in the nest for an unusually long time it could well be that she is feeling broody and, if she is still in the same nest at feeding time in the afternoon, she should be immediately placed in a broody coop. Three or 4 days in the coop should be sufficient to remove her broodiness and to prevent her reaching the full body heat which is part of its development. By taking this action, several days of egg-production can be saved. If the hen is allowed to sit on the nest for a longer period, it may take several days more to remove her desire to sit.

Merely keeping the hen off the nest is not always sufficient to prevent her from becoming broody as her body temperature also needs to be reduced without making any change in her diet. Therefore, it is always wise for the domestic poultry man to have a broody coop especially constructed for the treatment of such hens. Once the hens are back to normal they can be returned to their house.

Broody coops (Figure 11) have bars or slats across the bottom and have legs about 60-75 cm (24-30 in) high and a trough along the front for the food supply. If the general stock is on *ad lib* feeding, then the broody hen should be fed early in the morning, i.e. about 1 hour after sunrise and should have afternoon corn at the same time as the rest of the flock is fed. The broody coop should be placed in such a position that the occupant can see her companions at liberty in the pen. A water trough should also be fixed inside the coop and should be kept replenished.

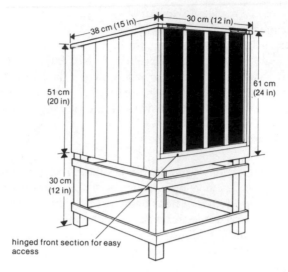

Figure 11. A broody coop. The slatted floor allows cold air to circulate around the broody hen thus reducing her temperature and discouraging her broodiness.

The object of having the slatted bottom to the coop and placing it on legs is to enable the cold air to circulate around the broody hen so that, when crouching down in the sitting position, she becomes cold and uncomfortable. As soon as the bird is seen standing up then the treatment can be assumed to be having the desired effect.

Chapter 5

Rearing

With the domestic poultry keeper, there will inevitably come a time
when the children will want to see some baby chicks. If there is no
cockerel running with the hens, you will have to buy in fertilised eggs
and, if one of your flock develops a broody instinct, i.e. stays put in
the nesting box, with or without eggs, this is an excellent opportunity
to start rearing chicks. Once you have a broody hen, you should
purchase, from a reliable source, preferably 12 but up to 15 fertilised
eggs. Bantams, which are the best brooders but smaller than the
standard hen, can sit on between 6 and 10 eggs. Although it is not a
difficult task to rear baby chicks, one should, of course, realise that
there is a right and a wrong way of preparing for and completing the
job.

Housing

If hatching is to be achieved naturally, it will be necessary to
accommodate the broody hen in a nesting box similar to that shown
in Figure 12. This type of nesting box is suitable not only for bantams

Figure 12. Nesting box suitable for bantams and standard hens.

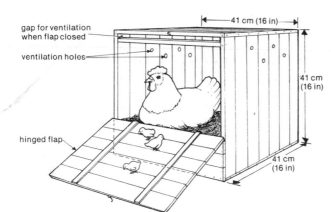

but also for the larger breeds. The breeding accommodation should be approximately 40×40 cm (16×16 in), with an opening flap at one end which should be hinged at the base and which can be secured at the top of the box when closed, leaving a 2.5 cm (1 in) gap at the top to allow for ventilation. At the base of the front, a retaining board 10 cm (4 in) high should also be fixed; this will help to keep the nesting material in place and will also keep the young chicks in the nest for the first few days of their existence. Suitable ventilation holes 1.25 cm (½ in) in diameter should be drilled at the back of the nesting box to allow excess heat to be expelled when the box is closed for the night.

Management of Sitting Hens

Broodiness is a most natural instinct in healthy hens and, even the so-called 'commercial' breeds, after they have laid a batch of eggs, will develop this instinct. Well-developed hens of the true breeds of poultry will make good sitters as well as good mothers, and the poultry keeper, providing he/she puts the birds under the right kind of conditions, will achieve good results. A nesting box (see above) should be provided, preferably inside a building or shed so that the sitting hen has added protection and can maintain the fever heat needed for developing the eggs which she is incubating. It is probably a good idea to place some china eggs in the nest for a day or two, just to make sure that the hen is really broody. If she continues to sit, the china eggs can be replaced with fertilised eggs when she leaves the nest to feed, preferably just before dark as the hen will remain on the nest during the hours of darkness.

Good management of sitting hens is most important throughout the process of incubation. During certain periods each day, the hen should be allowed enough time for feeding and exercise and scratching about — she should be kept off the nest for about 15 minutes each day. One important word of advice; do not allow the hen back onto the nest until she has produced some droppings otherwise she may foul the nest and may possibly infect the eggs with bacteria. If this situation should arise, then the nest and the hen should be cleaned at once.

The observant breeder will check the condition of the hen when she is on the nest. When she is fit and healthy, the eggs entrusted to her will have a glossy appearance because of the oil absorbed from her skin.

At the start of the incubation process, you will probably notice that the hen continues to sit tight on the nest, ignoring the desire to feed and drink. If this happens, you should lift her gently off the nest and place her down near food and water. The hen should not be left to return to the nest of her own accord or she will not be off the nest for a sufficient length of time. Some amateurs are inclined to make the mistake of allowing their hens to return to the nest too quickly; the eggs need to be aired for at least 15 minutes each day.

Hatching

When the eggs are about to hatch, the hen should not be disturbed or taken off the nest. This will disrupt her natural instincts and she must be allowed to remain on the nest. At this point, it would be as bad to remove the sitting hen from the eggs as it would be to open the drawer of an incubator because the eggs now have the correct amount of moisture and are at the correct humidity to allow the chicks to free themselves from the shells. If cold air is allowed to surround the eggs at this time, the inner membrane of the shell invariably becomes harder, thus making it very difficult for the young chicks to escape from the shell and start their new life in the outside world.

The rate of development of the embryo can vary but normally a healthy chick should be ready to break the shell at some time during the 21st day of incubation. As soon as the chicks begin to chip at the shells, the hen should be fed and watered without being taken off the

Figure 13. As soon as the chicks have hatched, and before their first meal, they should be placed in a specially-constructed rearing coop.

nest. Apart from this, it is best not to interfere. Normally, it will take between 6 and 10 hours for a chick to leave the shell completely. As a general rule, it is inadvisable to remove any of the chicks from the nest until the full hatching time has expired. Then, and only then, can the young chicks be removed to the rearing coop (Figure 13). If there is any undue delay in a chick emerging from the shell, then a small portion of the shell can be removed, just to make sure that the bird is still alive. This should be immediately obvious because a chick, if alive, will move as soon as it encounters fresh air. If everything seems to be progressing normally, it is best not to interfere.

If any eggs remain and show no signs of life, they should be removed. The hen should then be lifted off the nest for a good feed and transferred to the rearing coop. (This should have been prepared beforehand.) The chicks can then be placed under the hen and the coop closed up immediately. It is important to leave the chicks with the hen for at least 36 hours or so before the coop is opened up again as this will help to provide the warmth and comfort necessary for the chicks to assimilate the yolk provided by nature for their first meal.

Quality of Eggs

Since egg-production has become a profitable proposition, there has been a tendency for the domestic poultry keeper to lose interest in breeding offspring from the various types of poultry. The production of eggs for consumption is a fascinating hobby but the domestic poultry breeder should also be aware of the conditions necessary to ensure successful hatching.

While in eggs produced solely for consumption, the quality of the yolk is the most important consideration, it is the condition of the albumen — the white part of the egg — which controls the strength of the chick. Unless the albumen is of firm consistency and quality, the chick produced from the egg will either be weak or even dead — a situation which applies whether the egg has been home produced or purchased.

It is essential that the prospective breeder acquires the best quality fertilised eggs possible so that the embryos will develop into strong and healthy chicks. Scientists have proved, after carefully observing the condition of eggs produced by hens fed on different types of food, that the quality of the albumen, which is so vital to the hatching and reliability of the chick, is influenced by the feeding method and the quality of the food given to the hens.

A good many amateurs expect all fertilised eggs to hatch when maintained under the right conditions and they are unable to understand why fertile eggs fail to hatch. Anything which causes the developing embryo to come out of position will invariably kill the chick and prevent the egg from hatching and it is the quality of the albumen which is at fault in the majority of cases.

Care of Young Chicks

After the initial period of 36 hours or so, the hen can be allowed out of the coop to scratch about. It is not advisable to allow the hen to scratch about in the coop because, until the chicks are old enough to get out of her way, they may be trampled to death.

It is a mistake to rear chicks indoors. The idea that a constant temperature will encourage growth is not necessarily valid as, by taking the chicks off the ground which is their natural habitat, they may not thrive satisfactorily.

For the first 10-15 days, the chicks should ideally be fed on chick crumbs or pellets (crumbs being the cheaper of the feeds). These types of feed are available from pet shops, corn merchants or direct from a mill. You can, of course, make up your own feed but this may be a little more difficult. The best method of feeding is to purchase a balanced feed which will contain all the ingredients for healthy growth.

Whether the chicks have been artificially reared or bred under a mother hen, the best method of feeding is 'little and often'. Young chicks cannot digest whole grain and, moreover, the ordinary meal fed to the general stock birds is not suited to hens with chicks. The most suitable feed for both mother and chicks is balanced chick-crumbs. This is rich in vitamins and minerals and is much easier to digest.

If the young chicks are to grow satisfactorily, make sure that they always have ample food and water. Also, make sure that they feed frequently in order to encourage growth during the first few weeks. They should be kept free from draughts and disease and should have sufficient exercise. It is best to allow them to help themselves by adopting the *ad lib* feeding method. Chicks do look rather frail in the early days but they are in fact quite strong and, providing that they are kept warm and dry, will continue to grow and develop.

When the hen has hatched her chicks, she should be left to her own devices. She will brood them most carefully for the first few days and nights and, if a chick tries to leave her, she will get it back very quickly. Instinctively, she will not leave any chick on its own to become chilled; if chilling does occur, death might well result because the feed is not absorbed properly. If all remains well, then the chicks will have a good appetite when she takes them home for a meal.

Vermin Control

While the chicks are very young and are being kept in a rearing coop and run, they should be safeguarded from all types of vermin. Rats can be a major source of danger for they can kill and even eat the young birds. However, providing that no spaces exist around the side of the house and run, and that the coop is closed at dusk, then all should be well. Nowadays, of course, it is reasonably simple to eliminate vermin by using the appropriate chemical substances, which are generally available from chemists. If you do use rat poison, make sure that you do so wisely by placing it out of reach not only of the chicks but also of birds and domestic animals.

Avoiding Cramp in Young Chickens

As a rule, there are only two reasons for cramp developing in young chicks. Firstly, it is due to wrong food being given, i.e. one that is not easily digested and is deficient in the necessary vitamins and minerals. Although the young chicks require very little food, they must have a correctly balanced diet from the 3rd day of life onwards, otherwise their growth and development will be retarded. However, if any weakness should develop, this will be immediately noticeable by the lack of strength in their legs. Secondly, cramp can result from rearing the chicks early in the year, i.e. during the coldest months, and keeping them in confined spaces for long periods instead of allowing them to run and scratch about. Cramp often develops when the confined spaces are also draughty and damp.

If you do encounter the problem of leg weakness in young chicks, then steps should immediately be taken to remedy it and to prevent its occurrence in the rest of the flock. The quality of the food should be improved and the chicks should be moved to more suitable conditions. Cramp sometimes occurs in growing chicks as a result of poor circulation induced by the cold and dampness underfoot. This weakness of the legs will soon be noticed — the young birds will squat for long periods and eventually become helpless as their toes contract. If this situation does occur, do make sure that it is the result of cramp and not the dreaded curled toe disease, which is caused by incorrect feeding and poor accommodation. The best cure is a move

48

to more favourable conditions and gentle massaging of the legs of the affected birds.

Artificial Hatching

If none of your flock shows any signs of becoming broody, then the alternative is to hatch chicks in an incubator. There are several good, reasonably priced incubators available to the amateur, ranging in price from as little as £50 upwards and with a capacity to hatch about fifty eggs. Some of the latest plastic incubators have transparent tops so that the poultry man can view the baby chicks as they hatch. The initial cost is the only outlay, as maintenance and repairs are almost non-existent. If you intend to hatch chicks on a regular basis then the capital outlay for an incubator of proven quality will be an investment that will produce dividends over a short period.

Types of Incubators
The incubators available (see Figures 14-17), whether they be ancient or modern, operate on one of three different principles: a) hot water, b) ducted hot air or c) electric heating. For the amateur who wishes to hatch only two or three dozen eggs at a time, a small electrically heated type is probably best suited to his needs. Whichever machine is chosen, it must be emphasised that the correct humidity must be maintained at all times, in and around the incubator, as this is the substitute for the natural lubrication of the shells which occurs when the eggs are under the broody hen.

Figure 14. The top section of this incubator is moulded in one piece with four double-plate windows for observation. Circular depressions in the base hold water to maintain the humidity and the thermostatically controlled heating element is suspended from the lid to give even heat distribution. The egg tray will hold about 60 hens' eggs or 40-45 turkey eggs.

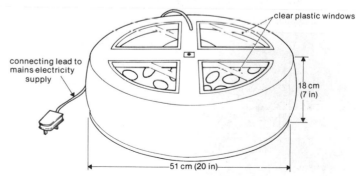

connecting lead to mains electricity supply

clear plastic windows

18 cm (7 in)

51 cm (20 in)

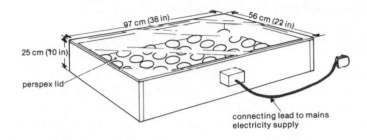

Figure 15. The perspex lid of this incubator allows a clear view of the eggs during incubation. The thermostat is easily operated and, although the heating element is 600 watts, it only uses 300 watts as it operates at ½ a cycle or less because of the sensitivity of the thermostat.

The incubator should be positioned in a well-ventilated, draught-proof and dry room or out-building. For the domestic poultry keeper there is no better place than a reasonably sized room in his own home, providing that this does not create a domestic problem! Hatching eggs require oxygen so there should be ample ventilation and the incubator must never be placed in a small confined space. This would result in a high percentage of chicks dead in their shells, because of the lack of fresh air during the last few days of incubation. To obtain good hatches, it is most important to follow the manufacturer's instructions at all times and these are generally quite simple. When setting up the incubator, care should be taken to make sure that it stands absolutely level so that a uniform temperature can be maintained.

Figure 16. This model incubator has a hen egg capacity of 80.

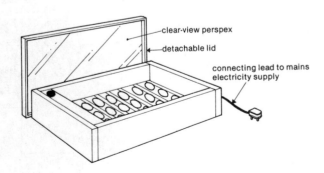

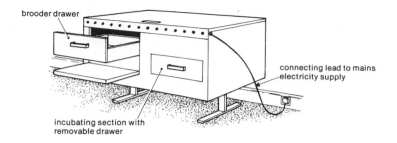

brooder drawer

connecting lead to mains
electricity supply

incubating section with
removable drawer

Figure 17. This electrically-operated cabinet-type incubator has a hen egg capacity of
100. An egg-testing light can be fitted if required. Once hatched, the chicks can be
placed in the brooder drawer for the first 36 hours.

As temperature control is the most important aspect of artificial
hatching, if there is no thermostat on the incubator, the room in
which it is situated should be at a constant temperature of 20°C
(70°F) for it is far better to have the correct room temperature than to
be continually regulating the heat in the incubator. It is also of
paramount importance that the incubator is not positioned near a
window, as the rays from the sun falling onto the machine can very
quickly increase the temperature inside, killing the embryos in their
shells. The temperature of the incubator for hatching needs to be
39.5°C (103°F). This can be monitored by a thermometer suspended
from the roof of the incubator and resting just above the eggs.
Nowadays, practically all incubators are thermostatically controlled
so there is no difficulty in keeping the temperature constant.

Positioning the Eggs
Once a constant temperature can be achieved, the fertilised eggs can
be placed in the incubator. They should not be placed haphazardly,
but positioned neatly, with even spacing between each egg, and then
left for the first 24 hours. After this time, the eggs should be turned.
As the broody hen sitting on the nest periodically changes her
position, and thus turns the eggs, so it is necessary to turn the eggs in
the incubator. Egg-turning is another important aspect of hatching, as
they should be evenly heated throughout. Unless turning is performed
on a regular basis, many of the embryos will die. This operation
should be done twice a day, i.e. morning and evening, and the best
way to check if the eggs are turned correctly, is to place a mark of
some kind on each side of the shell. For instance, they can be marked

'side 1' and 'side 2'. By ensuring that side 1 is uppermost in the morning and side 2 in the evening, you can ensure that the eggs are being turned correctly. If the eggs are placed on trays, these should also be turned round in case there is a variation in temperature within the incubator.

Cooling the Eggs

The successful hatching of fertilised eggs placed in an incubator, is dependent on the handling of the eggs during the incubation period and the control of temperature during the entire process. If too high a temperature is maintained, this may result in excessive loss of moisture content from the eggs or could possibly kill some of the embryos. During the first few days, the egg cooling process is automatically achieved when the incubator is opened for the turning of the eggs. When doing this, you should make sure that the temperature is not allowed to fall too rapidly — hence the need to maintain a room temperature of about 20°C (70°F) minimum. It is advisable to leave open a small window in the room to allow for adequate ventilation. A period of 2-3 minutes, or less, is ample time to allow for egg-turning and no additional cooling time should be allowed until after the first 8 days. Only then should the length of cooling time be extended and the period of extension will obviously depend on the room temperature. If the temperature of the room can be maintained at approximately 20°C (70°F), then a period of between 5 and 10 minutes will be sufficient additional cooling time. This is approximately the length of time for which the broody hen will leave the nest to feed and pass droppings before returning to her clutch of eggs. This airing of the eggs is another important process of hatching. There are no hard and fast rules governing the cooling period, because of the variation in ambient temperatures from day to day. Common sense therefore, must prevail.

Fertilisation Check or Candling

After the 8th day, it is advisable to check that each egg has been fertilised. It must be remembered, however, that the eggs must not be allowed to cool and that the incubator must not be opened too frequently, as the humidity which is so vital to hatching the eggs will be affected. Allowing cold air into the incubator causes the membrane within the shells to harden, placing the chicks in permanent confinement, as they are then unable to break through the shell when ready to hatch.

Testing, or 'candling' as it is called, can be carried out on the 8th day of incubation. By this time, the embryos in the fertilised egg will

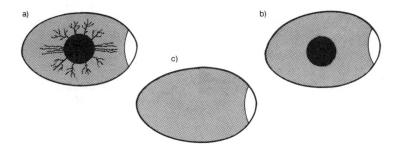

Figure 18. Appearance of eggs when held against a bright light ('candling'): (a) fertilised egg with viable embryo — note the blood vessels (b) fertilised egg with a dead embryo (c) infertile egg.

have grown to a noticeable size and, if the egg is placed over a bright light, a dark blob will be visible in the centre with small threadlike strands radiating from it; the air pocket should also be noticeable at the broad end of the egg. (Once the chick is fully developed and alive inside the shell, it obtains oxygen from this air pocket shortly before chipping commences.) An infertile egg will appear perfectly clear inside. By testing the eggs in this way (Figure 18), any infertile eggs can be removed from the incubator. If they are left, it is very likely that they will go bad and the resultant gases may possibly have an adverse effect on the other eggs. Therefore, although candling is not essential, it is advisable.

For the benefit of those readers who wish to examine their eggs in this way, the equipment illustrated in Figure 19 is all that is needed to complete the job quickly and efficiently.

Figure 19. A candling box used for testing incubated eggs for fertility. The egg is placed in the aperture and the strong light shining through it will show whether an embryo has been formed. The term 'candling' derives from the days when lighted candles were used.

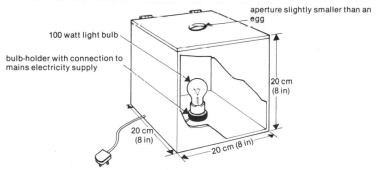

The Critical Period

Just as the broody hen will sit tight on her nest for the last 2 or 3 days of incubation, so it is necessary for the incubator to remain closed for this period. It cannot be over-stressed, therefore, that, from the 19th day, the incubator lid or drawer MUST NOT BE OPENED under any circumstances. If it is, the humidity level will be considerably reduced with disastrous results. This can be a trying time for the domestic poultry keeper but patience must prevail. For amateurs, to see chicks hatched under their very own supervision for the first time is an experience they will never forget.

Figure 20. Types of equipment used for brooding chicks: (a) infra-red brooder (b) hover brooder (c) brooder raised during daytime. The area surrounding the brooder should be enclosed to prevent the chicks from straying too far from the warmth of the lamp.

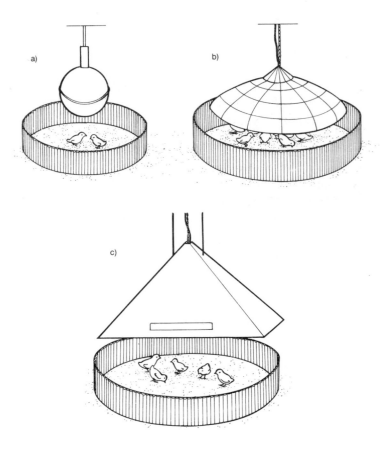

Rearing after Incubation

Well before the chicks are due to hatch, consideration must be given as to how they can be given the best start in life. As they have no broody or mother hen to nurse them through the first few vital days of their new existence, the newly hatched youngsters must be placed into an environment which is an appropriate substitute. It is necessary therefore to purchase, or construct, some type of brooder or brooding pen (Figure 20).

The brooding pen should be prepared a day or two before the chicks are due to hatch, so that the heating lamp can be positioned at the correct height to give the required amount of heat. It should be positioned in a suitable room or outbuilding, draught-free and with good ventilation.

The brooding pen illustrated in Figure 21 is one of the simplest and cheapest. It consists of a hanging electric lamp, preferably infra-red to avoid creating too bright a light, a chick-feeder or trough and a suitable water container. The lamp, or brooder, is suspended over the centre of the pen at a height which will maintain the required temperature of 32°C (90°F) for the first week, both day and night. The area should be surrounded with a suitable material to exclude draughts, e.g. hardboard or corrugated paper, approximately 30-60 cm (1-2 ft) high and enclosing an area of approximately 90 cm (3 ft) in diameter, depending upon the number of chicks hatched. After the first 7 or 8 days, the temperature can be gradually lowered to 26°C (80°F) for the following 7 to 8 weeks or until the chicks are fully

Figure 21. A typical brooder unit. The protective wall, of hardboard or corrugated cardboard, keeps the chicks within reach of the warmth. The lamp can be raised and lowered to adjust the temperature.

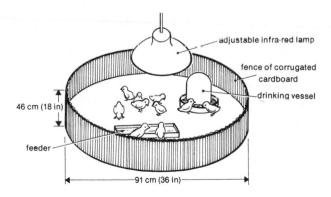

adjustable infra-red lamp

fence of corrugated cardboard

drinking vessel

46 cm (18 in)

feeder

91 cm (36 in)

feathered. The newly hatched chicks will need artificial heat until they have grown their true feathers, at which time they will have sufficient natural insulation to retain their body heat.

Do make sure that the electric lamp can be raised and lowered for, as the chicks develop their plumage during this 8-week period, they will gradually require less heat. You should, of course, check the baby chicks at least twice a day, so it will be immediately noticeable whether they require more or less heat. If the youngsters are crowding together directly underneath the lamp, then it is most likely that the lamp needs to be lowered to increase the temperature. On the other hand, if the young chicks are standing near to the outer rim of the breeding pen, then the lamp can be raised a fraction. Because of variations in the outside temperature, common sense inevitably plays a large part in determining whether the conditions are just right and you should not rely solely on the thermometer.

If the new additions to the flock have been allowed to develop normally, it is quite possible that the artificial heat can be removed completely after the 8th or 9th week. They can then be placed in their own rearing pen, so that they have access to the good mother earth.

Hatching and Rearing Bantams

Bantam fowl can be hatched and reared by exactly the same methods used for the standard breeds of poultry, i.e. either with the assistance of a broody hen (for bantams make good mothers) or with an incubator. There are a few smaller incubators now available to the general public, many with a clear-view cover, sometimes in the form of a dome; these permit constant supervision with the minimum of disturbance. These domestic incubators are quite reasonably priced and are a good investment, even when hatching only a dozen or so eggs.

One small difference between hatching and rearing bantams and the larger fowls is that of 'caring'. As the newly formed chicks are much smaller, it is recommended that a little extra care be taken during the initial stages, especially during the first 6-8 weeks after hatching. If you are scrupulous about meeting all the necessary conditions exactly, then success will be assured.

Hatching and Rearing Turkeys

Turkey eggs can be hatched in much the same way as the eggs of the ordinary domestic hen. The three main points are:

a) the size of the turkey egg which is much larger and therefore requires more space,
b) the temperature required to achieve a correct hatch (39.5°C/103°F)
c) the additional length of time needed for the young turkey poult to emerge from the shell.

Once the eggs to be hatched have been selected, which should be within a few days of collection, the remainder can then be used for human consumption — only top quality eggs should be used for hatching. The selected eggs should be free from any blemishes, well proportioned and with no cracks or dirty marks. Eggs which have hairline cracks can be used for hatching but the degree of hatchability may be somewhat reduced. If any doubt exists as to the quality of the eggs, the candling process (see p. 52) will reveal any hairline cracks. It is, in fact, always wise to use the candling method in the selection of eggs for hatching because any defects, such as blood spots on the yolk or watery albumen, will be noticed.

If incubation is to be achieved naturally, i.e. by using a broody turkey hen, it is as well to remember that the nesting box required needs to be, at the very least, half as large again as that used for the domestic hen because of the size difference between the birds. A standard tea-chest would be ideal for this purpose. Apart from this, the natural method for hatching turkeys is the same as for the standard fowl.

Because of the difference in the size of the eggs, incubation takes a little longer than for the standard domestic egg. Four weeks is the usual period required before the young poults are ready to hatch and it is on the 26th or 27th day that chipping will commence, after which it will generally take about 35 hours for the hatch to be completed. The eggs will not all hatch at the same time. Therefore, an extra day or two should be allowed for the entire hatch to be completed.

Once the poults are hatched, they can be moved, together with the mother bird, to the rearing coop, where mothering can be continued as for the standard domestic fowl. The only exception is that, because of their stupidity, turkey poults will not feed easily unless given a helping hand by the breeder. Usually, for the first day or two, because the young turkeys have difficulty in finding the feed and water troughs, it is necessary to watch over them and literally take them to where the food and water is situated.

If hatching is to be achieved artificially, i.e. in an incubator, then the same principles apply as for hatching the domestic hen's eggs (see p. 49). The most important points to remember are that an incubator holding, say, 50 or so domestic hen's eggs will hold only between 30 and 40 turkey eggs. Never allow the temperature above the eggs to exceed 39.5°C (103°F), as too high a temperature will increase the mortality rate as well as producing poor quality chicks. (This is assuming of course that all the eggs were fertile when placed in the incubator!)

To obtain maximum hatchability, the eggs should be turned regularly during the first 3 weeks of incubation — at least 3-4 times daily — a higher turning rate than that used for domestic hen's eggs. If the eggs are turned insufficiently, the embryos tend to become stuck to the shell membranes where they will undoubtedly die. Turning is not so important during the 4th week, as the young poults will be almost fully developed and there will be some natural movement within the shell.

Although candling can be carried out after the 7th day of incubation, it is probably better to wait until about the 18th or 20th day because, by this time, any dead embryos within the shell will be more easily recognised and the egg can be removed from the incubator. By this stage of development, a healthy embryo will almost fill the shell whereas, if the embryo has died, a large portion of the egg will remain clear.

Apart from the points mentioned above, incubation should be carried out in much the same way as for the standard domestic poultry egg.

Chapter 6

Disorders, Diseases and Moult

Although few of the diseases mentioned here will be encountered, it is as well to be aware of possible troubles.

Common Disorders

Bumblefoot

Bumblefoot is not a disease but a localised infection of the feet which sometimes occurs in adult fowls. This is usually caused, in the first instance, by a bruise or cut on the foot. Affected birds may appear to be lame and there will be swelling and inflammation in one or both feet. The swelling usually occurs between the toes and can get quite large in some cases. It is invariably caused by bacteria entering the foot through the bruise or injury. Sometimes the feet get cut and dirt accumulates in the wound, which becomes quite hard and overgrown with skin.

If a hen shows signs of bumblefoot, the affected foot should be bathed frequently in warm salt water. At the same time, attempts should be made to remove the hard dirt from the wound. When the swelling is large and completely covered by skin, lancing is probably the best solution. Make sure that all instruments are sterilised and afterwards wash the wound with antiseptic and cover it with sterilised material. Change the dressing frequently until the wound is healed.

If this problem is allowed to continue the sinews will become infected and egg-production will suffer.

Crop-Binding

Crop-binding is usually caused by careless feeding and the absence of grit, as well as by allowing the hens to eat long grass or hay. (Hay must never be given to fowls, either in their nests or in the pen.) it usually occurs when the hen is in full lay and, if treated at once, the blockage can soon be removed. The first signs are a bird which sits or mopes about looking sorry for itself, sometimes taking up food and putting it down again, and drinking large quantities of water. The hen will have a full crop even before the morning feed. The main cause of crop-binding is invariably the result of eating long grass, which becomes twisted and knotted inside the crop. After a time, the crop will feel hard and the entrance to the passage leading to the gizzard will become blocked. One of the remedies is to give the fowl 2 teaspoonfuls of salad oil and allow it a free supply of warm water. After a short time, gently rub the crop with the thumb and finger to try and move out some of the contents. If there are no favourable results after the treatment has been repeated two or three times, and if the crop is still hard, then place the hen in a dry pen for 24 hours with no food but with an ample supply of warm water. If all efforts fail then the only remedy left is to operate, i.e. to open the crop and remove the contents. It is not a difficult operation, but is best left to a veterinary surgeon.

Egg-Binding

Egg-binding is usually caused by a broken egg in the oviduct, which will invariably cause inflammation of the oviduct wall. This often happens after a hen has been frightened when, because of some sudden movement — flying into an object or jumping from a high perch — a shell-less egg breaks in the egg passage. An affected fowl may be seen standing erect with its tail down and head erect. The feather around the vent will usually be wet and the rest of the plumage will be ruffled. If the fowl is not dealt with quickly, then it will die in a day or two. It can also happen if the oviduct is too small to allow the smooth passage of the egg. Sometimes a broken egg in the oviduct can be the result of an over-active cockerel if one runs with the hens. The loss of birds through egg-binding could be avoided if poultry keepers understood how to treat the birds when the first symptoms appear. The birds should be watched closely and, if the cock or cockerel is seen to be over-active, then his time with the hens should be limited to shorter periods.

Some young pullets can die before they pass their first egg because of over-fatness of the egg-producing organs usually caused by a genetical problem. However, an experienced poultryman can soon tell by the look of the birds when something is amiss. Even an inexperienced poultry keeper should realise when a bird is not laying properly, if he sees the bird going to the nest several times without laying.

If this happens the bird should be caught and close examination will most likely reveal an egg near the vent. As soon as this is noticed, the bird should be held over a container filled with steaming hot water so that the steam can be allowed to permeate the entrance of the vent. The effect of the steam will be to soften the fatty tissue and make the area of the vent much more flexible, thus allowing the hen to pass the egg more easily. After this treatment the bird should be able to lay the egg within an hour or so.

This particular problem is due to the fact that the skins of the eggs are rather thin and hens which are at peak laying usually have two, or sometimes three, eggs in the oviduct which are nearly completed, except for the shell. Anything, therefore, which causes the hen to jump or jerk sharply, can result in the breaking of the skins. The fluid will ooze from the vent and, if the bird is not dealt with quickly, it will most likely die. Often, a hen will be found dead on the nest. Such deaths are invariably put down to egg-binding, although this is not generally the cause. It is more likely that the skin of the shell-less egg has remained inside the egg passage. The hen therefore thinks that she still has an egg to lay, which causes her to strain. This, in turn, will cause a rupture of part of the egg passage.

If the hen can be caught and examined, the skin of the egg may well be seen just inside the egg passage. If so, it can quite easily be removed by gentle pulling with a pair of tweezers. This must be done very carefully or the skin may break inside.

Egg-Eating

Egg-eating is another bad habit which often results from one or more hens laying soft-shelled eggs or eggs without a shell. It can also result from a bird dropping an egg from its perch. This may well be scrambled on the floor of the house and hens, being curious characters, will soon investigate the contents. Once they have tasted an egg, they will do so again and again. This problem is also encountered in hens kept in a confined space.

The formation of this bad habit can be avoided by allowing the hens free access to poultry grit, which can be purchased from either your local corn merchant or pet shop. 'A busy hen will give no trouble.'

When some of the birds are eating eggs and all other remedies have failed, the whole of the contents of an egg should be emptied and the shell filled with English mustard and cayenne pepper. The offending bird may eat the whole of it but certainly won't forget it in a hurry!

Feather-Eating

This is not a disease, but rather a bad habit started by one hen and soon copied by the rest of the flock. It is rather difficult to get rid of unless a remedy can be found. The problem of feather-eating is usually caused by hens being in a confined space with nothing to occupy them and so they get bored. This boredom leads to the hens pecking themselves and each other and, when they realise that they can also pull out the feathers, they think this is great fun. Before long, many of the birds will be bald about the rump and along the neck.

If this problem of feather-eating is noticed in time, it may often be remedied by allowing the hens a longer run and increasing their supply of green food — by offering brassica plants or stumps and also by throwing them weeds pulled up from the garden. Another solution is to obtain a supply of quassia, this is a strong solution which has a very bitter taste and can be daubed quite safely onto the feathers of the birds with no adverse effects. Also a few old Brussel sprout stalks can be suspended from the roof inside the hen house, just above the heads of the birds, so that they will have to jump to peck at them. This could be described as a form of occupational therapy!

Soft or Shell-less Eggs

Some hens will, on the odd occasion, lay eggs without shells; this is generally due to an insufficient supply of egg-forming materials and usually happens when hens are in full lay and is not exceptional. At this time, the hens can make eggs faster than nature can shell them. It has been thought impossible for hens to lay more than 1 egg in 24 hours, or rather to shell them in that time. However, this is not the case and I have experienced pullets of about 7 months laying more than 1 egg a day. Hens which lay shell-less eggs are not always out-of-sorts; it is just nature's way of saying that the hen is over-producing. Shell-forming material, in the form of broken oyster shell and flint dust (poultry grit) should be available to the birds at all times.

The production of double-yolked eggs weakens the egg-producing organs and is another cause of hens laying eggs without shells. Hens which are fat internally due to inactivity may also produce soft eggs.

Diseases

Practically all the diseases mentioned below are not usually encountered by the sensible domestic poultry keeper. Should any of them occur, however, then veterinary consultation will be necessary to obtain the appropriate antibiotics.

Air Sacculitis (Chronic Respiratory Disease)

Air sacculitis is a respiratory condition which can affect young chicks between the ages of 6 and 10 weeks, especially if they are intensively reared. This ailment can be expensive in terms of mortality, as well as through loss of weight and poor carcass quality, but it is rarely a problem for the domestic poultry keeper. This disease is caused by a virus.

Arthritis
This is principally a disease of young chickens and the predominant symptoms are lameness with some bacterial infection of the leg joints, particularly the hocks. This disease is commonly caused by infection with the staphylococcal bacteria.

Aspergillosis
A respiratory disease, this can be caused by mouldy straw, wood shavings or hay. Such bedding materials may harbour the fungus *Aspergillus*, the spores of which can be inhaled by the chickens. Affected birds exhibit respiratory symptoms including gasping.

Avian Encephalomyelitis (Epidemic Tremors)
This is a nervous disease of chicks up to 5 weeks of age. The causal virus is passed through the eggs, and, consequently, all breeding stock should be vaccinated against this condition. Chicks quiver and shake, then exhibit lack of coordination of the legs, followed by general paralysis and death.

Blepharitis
This inflammation of the eyelids may be due to a number of factors, e.g. nutritional deficiency or dusty conditions.

Coccidiosis
This is caused by protozoan parasites, *Eimeria*, invading the caeca and intestines and causing damage to the cell walls of the gut. Control is by a coccidiostat in the feed and good litter conditions. Treatment is by proprietary drugs in the water. If the condition is not serious, then a good dose of Epsom salts added to 1 pint of water and put down for the hens to drink, may well eliminate the condition.

Colds
As in human beings, these are bacterial or viral respiratory infections involving the eyes, nose and associated sinuses. In poultry, they frequently result from long periods of confinement in chick boxes or crates, but may be due to feeding a ration which is deficient in Vitamin A.

Crazy Chick Disease (Nutritional Encephalomalacia)

Young chicks suffering from this disease tumble over backwards, fall on their sides and do bicycling movements; a number die. It is usually caused by a deficiency or destruction of Vitamin E in the breeder or chick rations, which causes brain damage. The symptoms are similar to Avian Encephalomyelitis (Epidemic Tremors).

Diarrhoea

Scouring is a common problem which may be caused by eating stale or mouldy feed and is associated with many diseases; including coliform infections, salmonellosis, fowl pest, fowl typhoid and semi-impactions of the gizzard.

Enteritis

Inflammation of the intestines by *Escherichia coli, Salmonella* etc. Intestinal parasites and excessive flint grit are predisposing factors. Treatment is by antibiotics.

Epidemic Tremors see Avian Encephalomyelitis

Escherichia coli Septicaemia (Coliform Infections)

This blood poisoning results from infection with *Escherichia coli* bacteria.

Fowl Cholera

This infectious disease is capable of causing severe losses in all species, but is most common in turkeys. A vaccine can be prepared to counteract this disease which manifests itself as colds, abscesses on the wattles or, more typically, by severe septicaemia and heavy mortality. Treatments include sulpha drugs and antibiotics.

Fowl Pest see Newcastle Disease

Fowl Pox

An infectious and contagious disease caused by a virus, this is characterised by growths on the head and in the mouth. Vaccination of growing stock should be carried out in areas where pox is prevalent.

Fowl Typhoid

Caused by the organism *Salmonella gallinarum*, this disease is spread by the droppings of infected birds contaminating foods, litter etc.

Vaccines for protection and a blood test for detection are available.

Gumbroro Disease (Infectious Bursal Disease)
This virus disease is mainly confined to broilers and growing pullets in the first 2-5 weeks of life. Characteristically, the bursa of Fabricius is enlarged. Mortality is often low but can be as high as 30 per cent. Certain farms or areas have a history of the disease. It is known to interfere with the development of immunity to other diseases. Vaccines are available for treatment.

Haemophilus Infection
This cold-like bacterial respiratory infection involving the eyes, nose and associated sinuses is due to the organism, *Haemophilus*. It frequently occurs after long periods of confinement in chick boxes or crates, but may be due to feeding a ration deficient in Vitamin A.

Haemorrhagic Disease
A sudden attack may occur about a fortnight after the birds have been treated for coccidiosis with a sulpha drug. Diets low in Vitamin K appear to favour the onset of this disease.

Infectious Bronchitis
A respiratory infection caused by a virus, the symptoms are sneezing, coughing and wheezing. Chicks will sneeze and have difficulty in breathing. In layers, egg-production falls sharply and abnormally shaped eggs may result.

Infectious Laryngotracheitis
This is an infectious disease of pullets caused by a virus. If severe, it is characterised by the coughing up of blood and high mortality. It is especially common in Lancashire and the surrounding counties. A vaccine is available.

Infectious Sinusitis
This inflammation of the sinuses and, in some cases, the lower respiratory tract, is caused by inhaling or digesting *Mycoplasma gallisepticum*.

Infectious Synovitis
Arthritis and synovitis caused by *Mycoplasma synoviae*. Lameness, swollen joints and breast blisters are symptoms of this disease. It can be responsible for considerable down-grading of the carcasses.

Marek's Disease

A major killer of fowls, this disease is commonest in growing birds and young adults. Affected birds are often recognised by paralysis of the legs and wings. Mortality can be severe. Protection is given by vaccination of 1-day-old chicks, preferably at the hatchery.

Mites

These minute parasites (Figure 22) may be found on the fowl and are generally associated with dirty conditions. They cause irritation and, in severe cases, anaemia and mortality. Thorough disinfestation of birds and housing is necessary to kill these pests.

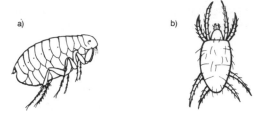

Figure 22. Parasites: (a) flea (b) mite. These will inevitably appear at some time or another, but if the hen house is cleaned and disinfected at reasonably frequent intervals, they will be effectively controlled.

Mycoplasmosis

This complex respiratory disease is caused primarily by *Mycoplasma gallisepticum*. Complications can be caused by a secondary infection of *Escherichia coli*. The virus may be controlled by vaccination.

Newcastle Disease

This infectious viral disease is also known as fowl pest. Nervous respiratory symptoms and diarrhoea with a high mortality rate are indicative of this notifiable disease. Vaccination as a preventive should be a routine operation with all stock.

Omphalitis

Bacterial infection of the tissues surrounding the vents of young chicks. These tissues become moist and putrefied.

Pullets' Disease

Of unknown cause, this is also known as 'blue comb' or avian monocytosis. Affected birds lose their appetite and often develop watery or whitish diarrhoea with a concomitant fall in egg production.

Pullorum Disease
Caused by *Salmonella pullorum*. Also known as BWD (bacillary white diarrhoea). Hatcheries carry out stringent hygiene programmes and a blood agglutination test on breeding stock to eliminate the causal bacterium. Affected birds have a hump-backed appearance, ruffled feathers, pasted vent and cheep continuously. High mortality rates can be expected. Lameness in older chicks may be due to this infection. It is rarely seen today.

Rickets
This weakening of the bones of young chicks can be eliminated by maintaining the correct levels of calcium, phosphorus and Vitamin D in the ration.

Salmonellosis
This is a group of diseases caused by bacteria belonging to the *Salmonella* group. There are no specific symptoms. Mortality occurs only in young chicks. It is sometimes transmitted via the egg or passed to the chick by dirty shells contaminated with *Salmonella*. It can be carried by adult hens which show no symptoms. The use of naturally clean breeding eggs and good hygiene is most essential to control this disease. It is now a notifiable disease under the salmonellosis order.

Stress
This condition of anxiety or 'one degree under' among a flock may be brought about by poor management, disease, disturbance, vaccination etc. It occasionally results in panic with subsequent deaths from suffocation.

Vitamin A Deficiency
A deficiency of Vitamin A, also known as nutritional roup, may cause retarded growth, weakness, lack of coordination and ruffled feathers. There may be a discharge from the eyes and nose. Sporadic outbreaks of Vitamin A deficiency occur, either because of inadequate diet or because of oxidation of the vitamins in the feed. This is likely to happen if the food becomes stale or rancid through being stored for too long.

Vitamin B2 Deficiency

This deficiency causes a condition known as 'curled toe paralysis'. Affected birds are reluctant to move, crouch on their hocks and, when disturbed, stagger forward on the outer sides of their inwardly turned toes. In the final stages, ailing chicks lie on their sides with their legs sprawled out.

Worms

Caecal Worms. Generally harmless, except that they harbour the parasite causing blackhead in turkeys.

Gape Worms: Give rise to gasping, coughing and signs of suffocation among chicks and poults.

Hair Worms: Parasites of the intestines which may produce severe and rapid emaciation in growers and, in layers, a steady fall in egg production.

Large Round Worms: Intestinal parasites.

Tape Worms: Intestinal parasites, which require an immediate host, usually a slug, snail or insect.

The Moult

Most hybrid breeds will, under good management, lay for up to 14 months before the onset of the first moult. The moulting period can entail up to 3 months with a considerably reduced egg-production or none at all. To avoid this, the majority of poultry keepers replace their stock at the completion of the first laying cycle. However, for the newcomer who intends to keep his flock for a second laying period, the following information may be of some help indicating when and how to cope with the moult.

Loss of egg-production commonly happens when the hens start their moult, usually at the end or near to the end of their first 14-month laying cycle. Some hens will have only a partial loss of feathers and others will literally lose all their feathers — a not uncommon occurrence. A well managed healthy hen which has completed her first laying cycle will not take more than about 6 to 7 weeks to change her feathers, whereas hens which are older will take 2 or even 3 months to complete the change.

If the younger hen starts moulting e.g. in the first week of July, she should have finished the moult by the end of August. This will give her time to revitalise herself before the onset of the winter months and she should start laying again at some time in October.

During the first part of the moult, the supply of food should be of a poor quality and feeding should be restricted. This will encourage the feathers to fall as quickly as possible. To try and shorten the period of the moult, some commercial breeders deprive the hens of food and water for a period of 24 hours or so — certainly not more than 24 hours. As soon as there are signs that the new feathers are being formed, change back immediately to the concentrated layer's mash mentioned on p. 29.

Apart from the above, there is nothing very special to note about the moult. It is only necessary to make certain that the birds are warm at night and free from draughts. Mismanagement, such as lack of water and/or feed, causes premature moulting and low egg production.

Partial Moult

Sometimes the newcomer to poultry keeping can be puzzled by a partial moult, i.e. when the hen shows a loss of feathers from the underside of the neck, but it very seldom lasts longer than 3-4 weeks. Nevertheless, it is sufficient to affect the egg supply adversely for a few weeks. The partial moult is nature's way of saying that the hens are laying an above average number of eggs. This may happen, for instance, then a breeder uses artificial lighting during the darker months to make the hens work overtime. It can also be experienced by a beginner using a high protein feed for a long period. The birds which are least likely to be affected by partial moult are those born in March, providing, of course, that they have been allowed to run freely and develop naturally.

Chapter 7

Preparing Poultry for the Table

Killing

One of the most efficient ways of culling chickens is by dislocation of the neck. When this is carried out correctly and swiftly there is no more humane a method. The person who is culling the hens should hold the bird by both legs and gather the main feathers of the bird's wings (the primaries) in the same hand (Figure 23). By adopting this grip, the bird will be unable to wriggle about. When held in this way, the back should be facing outward. The bird should then be laid across the left knee and the head taken between the first and second finger of the left hand (*vice versa* for left-handed people). The comb should be lying in the palm of the left hand and the fingers should be closed about the neck immediately behind the head, i.e. making a clenched fist. The neck is then drawn to its full length by the left hand. At the same time, the head should be thrown sharply backwards by a not too vigorous tug, which will break the vertebral column. The neck must be pulled fully out immediately behind the head so that the veins and nerve cord are completely severed. This

Figure 23. The correct way to cull a hen. Dislocation of the neck is achieved by a quick movement of the wrist, combined with the application of downward pressure.

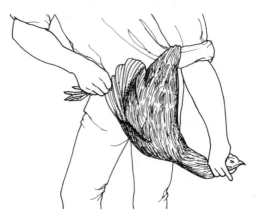

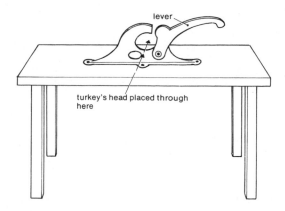

Figure 24. This is one of the most humane turkey-killers available and has the approval of the RSPCA and appropriate Government department. The head of the bird is slotted into the arc so that it protrudes slightly on the other side. When the lever is pulled, the head is dislocated from the neck and the main artery is severed. Death is instantaneous and the bird feels no pain.

method ensures that the bird experiences virtually no pain at all because, once the spinal cord is broken, no messages can be passed to the brain and all feeling in the body below the break is lost. When properly done, a process which should take not more than 2 seconds, there should be a break in the column of the neck of not more than 2.5-4 cm (1-1½ in) and the head should be connected to the neck only by the outer skin. Under no circumstances should the skin be torn in any way as this tends to stain the area with blood.

Turkeys are also most commonly killed by dislocation of the neck. Again, make sure that the main artery and the spinal cord are separated from the head so that a suitable space is left into which the blood can drain. If the blood is not drained properly, a slight reddish colour will be left in the flesh. After culling, it is advisable to hang the turkey up by its feet so that the blood drains more quickly.

Smaller turkeys can be killed in the same way as standard hens and just as easily. With larger turkeys, however, this is not quite so simple, as few people have either the reach or the strength to complete the operation satisfactorily. This problem can be overcome by calling on the services of a friend or neighbour who will be required to assist in one of two ways.

In the first method, one person should stand on a firm raised object and hold the turkey, head downwards, by the legs and wings, using both hands. The other person should place a broom handle or an iron

71

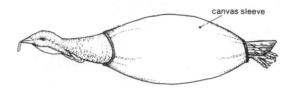

canvas sleeve

Figure 25. For ease of handling when culling turkeys, the bird may be placed in a canvas sleeve with the head and neck, and the feet, protruding.

bar over the head of the bird and stand with one foot on each end while, at the same time, the turkey is pulled quickly and firmly by the legs until the head is dislocated from the neck and the spinal cord and blood vessels are broken. Alternatively, a V-shaped piece of wood can be fixed to a suitable surface; the head of the turkey is slotted into this and the bird is held firmly in position. The assistant then immediately pulls the legs sharply to dislocate the neck.

It is possible to purchase a hand-operated 'killing machine' which is quite a simple device and suitable for one-person operation (see

Figure 26. Turkey culling machine based on an original design by E.S. Cockram (Patent Pending) and approved by appropriate authorities. The bird is placed in the canvas sleeve and the head is passed through the aperture at the lower end and slotted between and just below the two metal arms. The arms are held firmly and close together while being pushed downwards. This action dislocates the head from the neck and the bird feels no pain.

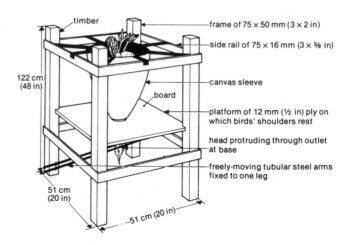

timber

frame of 75 × 50 mm (3 × 2 in)

side rail of 75 × 16 mm (3 × ⅝ in)

122 cm (48 in)

canvas sleeve

board

platform of 12 mm (½ in) ply on which birds' shoulders rest

head protruding through outlet at base

freely-moving tubular steel arms fixed to one leg

51 cm (20 in)

51 cm (20 in)

Figure 24). The turkey is placed in a canvas sleeve (Figure 25) from which the head protrudes. The head and neck are placed into the slot in the machine and the operator pulls the lever, dislocating the bird's neck. There are also more elaborate machines, such as the one shown in Figure 26, which are mainly used by the commercial turkey farmer who has a hundred or more birds to kill in a short time. For the small-scale breeder, such machines represent an unnecessary expense.

Plucking

Once a bird has been culled, it is much easier and quicker to pluck it while it is warm because the skin is still very supple and the feathers come out easily. The best way to pluck a bird is to sit on a short-legged stool or low chair, holding the bird by the legs and resting it on your knees with the head dangling downward. It is usually best to start at the back of the neck and work your way towards the tail, making sure that the feathers are plucked in the reverse direction to which they lie on the body (Figure 27). No more than three or four feathers should be plucked at any time or the flesh may tear.

Start at the back of the neck by first taking a few feathers between the thumb and forefinger and then giving a sharp tug towards the head, i.e. away from the lay of the feathers. The art of plucking is to pull the feathers sharply, as this will avoid tearing the skin. After the

Figure 27. Plucking a hen.

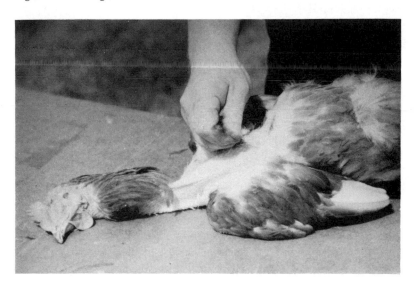

back of the bird has been plucked, turn the hen over and continue to pluck the breast and legs etc. in the same manner. The larger feathers on the wings, which are known as primaries, may be more difficult to extract; it is best, therefore, to pull them out straight, one at a time. Continue to pluck the neck up to about 5 cm (2 in) from the head, always leaving some feathers at the top part of the neck. The method used for the large wing feathers should also be applied to the tail feathers.

Wet Plucking

This is an alternative method which is particularly effective if the carcass has been allowed to cool. If a bird has been killed on the day previous to plucking and the carcass has become cold, immerse it in clean water which has been heated to a temperature of 54-60°C (135-140°F) — a higher temperature would have the effect of cooking the outer skin. Allow it to stand for approximately 1 minute at this temperature, after which it should be removed from the water and plucking can be commenced. The feathers will almost fall out and, because they are wet, will not be blown about.

Preparing Poultry for the Oven

Once the bird has been plucked, it must now be prepared for the oven. The utensils needed for this operation are: a sharp vegetable knife, a plastic bag or newspaper and a cloth or absorbent paper for wiping the hands.

Lay the bird on a work surface on its breastbone and, with the finger and thumb, lift the skin at the base of the neck and cut it, making a 2.5 cm (1 in) incision. Then place the tip of the knife under the skin and make an incision along the neck until the remains of the feathers on the head are reached (Figure 28). At this point, hold the upper part of the neck firmly in one hand and, with the other hand, pull the neck out from the severed skin. Cut off the head and upper part of the skin at the point just below where the blood has coagulated. You can now make a circular cut around the base of the neck and then twist at this point until the neck separates from the body. Once the head and neck have been removed, turn the bird onto its back. Using the finger and thumb, take a grip of the flesh halfway between the breastbone and the vent. With a sharp knife, cut horizontally into the flesh and layer of fat to a depth of about 2.5-4 cm (1-1½ in), taking care not to cut the intestine. You can then get two or three fingers inside the carcass and, by grasping the stomach or gizzard

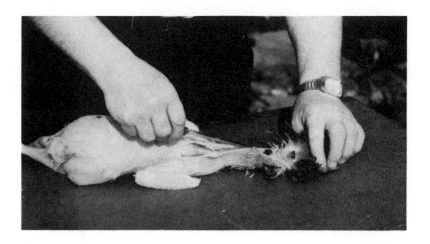

Figure 28. Preparing a hen for the table.

with two fingers, you can draw out all the entrails. Once the stomach, intestines and liver have been removed, you can place your hand deeper inside the bird and extract the heart and lungs. Of these parts of the bird, you should save the heart, liver, the outer case of the gizzard and, of course, the neck; these are known as the giblets. If you wish to use the liver, you should be careful not to break the small greenish gall bladder, the contents of which will not only discolour the liver but also taint the flavour. Should any of the organs break or tear, wipe up the mess with a damp cloth or absorbent paper.

If the bird is to be frozen, first rinse the inside with cold running water then drain the carcass and place it in the freezer in a plastic bag. (It is, however, considered unwise to wet the carcass if the bird is to remain in a raw state for a day or two.) Hens should not be stuffed before freezing as this limits the storage life of the birds. A stuffed chicken should be consumed within 1 month. The giblets can be stored inside the bird providing that they are wrapped separately. Quick freezing should be carried out at temperatures of −20°C (−4°F) or below so that the natural quality and flavour are preserved.

One of the major problems of deep-freezing poultry is the formation of ice crystals in the fibres of the flesh. This takes place at between −1°C and −4°C (30°F and 25°F). Once the produce has reached a temperature of −18°C (0°F), or below, ice crystals can no longer form; therefore the faster the food is frozen, the better chance there is of retaining the original flavour and tenderness.

Chapter 8

Poultry Manure and Its Uses

Manure is a term that is applied to all types of plant fertilisers, whether in a liquid or solid state. In the first instance, it is perhaps best to explain in general terms the different types of manure available and their relative values. There are two groups of manure: organic and inorganic. Organic manure is essentially a natural product, consisting of poultry and all farmyard manure compost. Inorganic manures are those which consist solely of pure chemical compounds, obtained by a manufacturing process or from crushed minerals.

Natural organic manures are by far the best and should be included in the compost heap, together with straw and other garden refuse; when rotted down and dug in, they add humus to the soil and put good heart into the land. Artificial or inorganic manures, in a form readily available as plant food, are to the land what tonics are to the human race. They act as a stimulant, but only for a short period, and are soon washed away by the rains, thus having no lasting benefit, to say nothing of their expense. Poultry compost, in comparison, will continue to benefit the soil for at least a good season, or even longer.

All manures have two main functions: firstly, they contain ingredients which can be used directly by the plants and, secondly (and this is particularly true of all forms of farmyard manure, including poultry manure), the process of fermentation and decay which takes place within the material, when mixed with the soil, releases many trace elements, e.g. iron and magnesium, from the soil which were not previously in an assimilable form. These elements are essential to the strong healthy growth required for a good crop. Also, because of the gases produced during the process of fermentation, the soil is kept open and its texture is lightened.

There is considerable nutritional value in a compost composed of poultry manure, lawn cuttings, leaves and other vegetable refuse. Cow manure, pig manure, wood ashes and weathered soot are also quite useful, because of the large amounts of potash and nitrogen which they contain. If fruit, flowers or vegetables are to be grown year in and year out on the same ground there are three main elements that should be added to the soil in the form of manure. These are:

a) nitrogen — for strong healthy leaf growth.
b) phosphorus — to promote good root and stem formation so that the plant can take in the plentiful supply of food available.
c) potassium — to encourage fruit and flower production.

It is not sufficient, however, just to add these to the soil; they must be added in such a form that they can be readily absorbed by the fine rootlets of the growing plants. Therefore, when poultry or farmyard manure is added to the vegetable or flower garden, time must be allowed for the manure to ferment and decay, so that the necessary elements can be released before planting takes place. This action should be completed in approximately 3-4 months.

Poultry manure, when mixed with straw, or better still, farmyard manure in reasonable quantitites, improves almost any soil, whether it be heavy or light. It is more beneficial than general farmyard manure because it is very rich in nitrogen, phosphorus and potassium. The proportion of nitrogen in the excreta of the poultry fowl is far greater than that of any other animal; in fact, the average quantities of these three main elements are, by weight, approximately 3 per cent nitrogen, 3.5 per cent phosphorus and 2 per cent potassium, the remainder consisting of minerals and other trace elements and vegetable waste. Because of the richness of poultry manure it is always advantageous to mix the fresh excreta with other forms of compost material, as the high nitrogen content accelerates its decomposition while at the same time adding all the other necessary ingredients to produce an ideal compost mixture. If you intend to construct a compost heap using poultry manure, allowances should be made to combat any loss of nitrogen which may be incurred by heavy rains. As much as 50-60 per cent of the nitrogen may be lost if the compost heap becomes sodden and, therefore, a wise gardener will always make sure that his compost heap is covered most of the time, so that heavy rains do not erode all the goodness.

Poultry manure is undoubtedly a valuable asset to the keen gardener and can be a profitable by-product from keeping domestic poultry — and one which entails no cost. The average layman may well have been led to believe that the white part of the birds' excreta consists of lime, but, in fact, it is this white deposit that contains all the nitrogenous goodness. This is highly beneficial to brassicas and many other plants. When incorporated with other materials, it forms a slow-acting fertiliser and is therefore best applied when the soil is being dug over. It can be applied to any type of soil, whether it be heavy clay or light loam. A good dressing should be applied during the autumn months. In the springtime, it may be sprinkled around growing crops, but you should make sure that the manure does not come into direct contact with the plants and also it should never, never come into contact with the young tender roots. The rate of the application of pure poultry manure should be in the region of about half a spadeful to the square yard (or square metre).

The plants to which poultry manure is best suited are: beans (all types), broccoli, Brussels sprouts, cabbage, carrot, cauliflower, celery, chicory, cucumber, leeks, lettuce, marrow, parsnip, peas, potatoes, radish, savoy, shallot, spinach, swede, tomato and turnip.

One important point to remember is that poultry manure, or for that matter farmyard manure, should never be applied when the soil is deficient in lime as the nitrates, phosphates and potash supplied by the manures will then be quite inaccessible to the plants. It is the lime in the soil which assists the bacteria in breaking down the organic matter, thus helping to release the essential plant foods and making them readily assimilable by the young plants.

Chapter 9

Poultry Keeping Month by Month

It should be borne in mind that the following descriptions of month-by-month activities apply to the northern hemisphere, particularly Britain, and that appropriate adjustments should be made by readers in the southern hemisphere.

January

Both breeding stock and layers need to be kept busy duing the short dull daylight hours which we experience at this time of the year. By scattering some grain, either in the hen house or in the pen, according to the prevailing weather conditions, the birds will be provided with the exercise and occupational therapy so vital in keeping them warm and active — an important factor for both good health and good egg development. If the birds are allowed to remain inactive for long periods then they run the risk of developing the habit of feather-eating. The birds will prefer grain for their afternoon feed. The feed itself and the activity involved in feeding will keep them warm if the weather is very cold and will enable them to settle down for the night much more easily.

January, surprisingly, is when serious hatching should start. All equipment, such as incubators, 'foster-mothers' and brooder equipment should be checked for cleanliness as this is absolutely essential. Whatever type of equipment is used, it should be warmed up for a few days beforehand so that the correct temperature and humidity are achieved. The young birds need to be kept at a steady temperature once they have been transferred from the incubator to the brooder.

This is the month when some domestic poultry keepers are lucky enough to find one or two broody hens in their small flock and it is an ideal time for hatching and fostering replacement chicks. It is no good finding a broody hen as late as June or July because chicks hatched during this period will not commence laying until November or December and, as this is the period of the year that has the shortest days, egg production will be at its lowest.

February

During the inclement days of February, it is desirable to keep the stock indoors so that it is not adversely affected by the wet, cold ground. Some people believe that stock which has ranged out of doors produces young chicks which are hardier and more satisfactory than those produced by more protected birds but this is a matter of personal opinion.

If an incubator is used during the early part of the year, a regular check should be made on the temperature of the place where the incubator is accommodated. A temperature of 20°C (70°F) should be maintained at all times for the most beneficial results.

For early rearing, the use of a small portable run which can be turned according to the wind direction is an advantage as, from the time the chicks are a week old, they can be allowed to decide for themselves the amount of time they spend running out of doors.

March

All poultry keepers look forward to March because, in this month, the days lengthen, providing longer evenings in which to do the essential work of looking after the young stock.

During March, which is frequently a wet and windy month, the domestic poultry keeper should ensure that the birds are able to seek protection from adverse weather conditions. This, and I make no excuse for mentioning it again, is vital to the maintenance of the birds' fitness and these sort of weather conditions can greatly reduce egg-production.

Heat reserves are lost much more quickly by the birds encountering high winds than extreme cold and, as egg development depends on the presence of heat reserves, care should always be taken to supply the best possible facilities. The period of between 6 and 8 weeks of age is a critical one for young chicks and so additional comfort should be provided in the form of heating whenever it becomes necessary.

On the day that the chicks are to be placed in their own quarters, the food and water troughs should already be in position inside the house and should be left there until the following morning. Next day, when releasing the birds into the run, the food and water should also be transferred. For the next 12 days, the birds should be confined to the run; after this period, they may be released on free-range. This will ensure that they return to their own house each night to roost.

April

If the domestic poultry keeper has chicks which are just a week old at this time, he should be aware that they generate a good deal of heat during the hours of darkness and care must be taken to see that they do not suffer from overheating. As a rough guide, a temperature of 24-30°C (75-85°F) will suffice, especially when the weather is favourable. It is interesting to know that 32°C (90°F) is recommended for all chicks during their first week; this should be lowered to 30°C (85°F) for the second week, after which the birds do not need a temperature of more than 24°C (75°F).

The most profitable period for the domestic poultry keeper is during the warmer months of the year, i.e. April to September. Therefore it is sensible not to dispose of hens during the months of May or June, despite the usually high prices available. Apart from the fact that the summer months are the peak laying period, the cost of feeding is reduced as the birds require less food to produce the same number of eggs which they laid during the winter months. Although they require less food, the birds must still receive the same sort of food, either a balanced layers' mash or pellets. Some people think that anything is good enough but this is an unfortunate mistake. Whether or not the birds are laying, they must be fed correctly; it is easier to put them off laying than it is to get them back on again and egg-production depends on the constant maintenance of the birds' heat reserves. Correct feeding is as important in April as it is in November.

May

There was, for a time, a tendency to regard later-hatched chickens as being of less value than those hatched in January and February. The fact is that, in many cases, pullets hatched in May and June, providing they are from reliable strains, are just as suitable as long as they receive the necessary attention during the rearing process; moreover they will prove to be just as reliable and profitable. The difference is that birds hatched in June and July will produce fewer eggs, as they will not start laying until the winter months. However, to regard any birds as being of no value, no matter what their hatching date may be, is unjustified.

The youngsters should not be put onto ground which has recently been occupied by a previous brood. Fresh land is needed to reduce the likelihood of infection and also to ensure a regulated fast growth.

June

When a few dozen hens are kept, June is to the breeder what March is to the housewife, the time for house-cleaning. Each house that is occupied must be thoroughly cleaned and it is useful at this time to have a spare house available as temporary accommodation for the stock. This will leave one house vacant for cleaning. After sweeping down the inside walls, remove the deep litter to the compost heap. After removing the droppings board and cleaning it and the perches, any parts which need replacing or fixing can be dealt with. The hen house should then be creosoted inside and out. For economical reasons, it is better to apply the creosote with a stiff brush of reasonable size as the bristles will penetrate the crevices in the wood more easily. A golden-coloured creosote should be used for the interior of the house and the heavy, darker type should be applied to the outside. Never use a spray to apply the creosote as this can be rather wasteful. After creosoting, leave all the doors and windows open so that the house can dry thoroughly. If the hens are kept in a pen, the ground can be dug over, wholly or partly, to sweeten it.

Never allow the birds access to the mowings from long grass as hens are not intelligent enough to realise that tough old grass is dangerous and, when eaten, may cause crop-binding.

Providing the same attention is given to the June broods as to those reared earlier in the year, they will give only half the trouble of those hatched in January and February yet still prove to be consistent layers.

From this time onwards, throughout the summer months, the hens will require more drinking water than usual because of the heat. Therefore, this should be checked morning and evening. Also, egg-eating may become more prevalent. One of the most effective ways of eliminating the problem is to give the birds an unlimited supply of empty egg shells. After feeding well on these they will lose interest in any other eggs.

On no account should the feed quantities during the summer months be reduced, because egg-production depends on the hens being uniformly fed; the number of eggs laid during the autumn will depend on the stock receiving good feeding during June, July and August.

July

There should be no change in the type of food given during the summer months as the birds should be getting plenty of nourishment to cover both a good supply of eggs and feather development. Forcing egg-production at this stage, or at any other time of the year, is poor husbandry and most unwise. Forcing hens to lay by consistently using high protein feeds can only have an adverse effect on their constitution in the long term and will only delay the inevitable rapid fall-off in egg-production at the onset of the moult. On average, a hen will lay only 5 eggs per week; trying to force an increase may well result in some of them laying soft or shell-less eggs. These can be broken in the nest and thus start the bad habit of egg-eating.

The number of eggs that each hen lays per week is influenced by the quality of the diet and as long as the digestive system is healthy and the food is being processed correctly, there is no possibility of the condition known as over-fatness developing in the birds. Occasionally during hot weather the birds will appear disinterested in their food; there is no need for anxiety when this is noticed. The only thing to do is to make sure that they have plenty of fresh water to drink.

August

It is sometimes believed that, if a healthy cock or cockerel, not necessarily over-active, is allowed to run with the hens, then egg-production will increase appreciably. This is no longer considered significant as the hens will continue to lay at their normal rate if fed correctly. The only difference is that, without the cock or cockerel, the eggs will not be fertilised and therefore cannot be used for breeding purposes.

To wait until the birds are in full lay before introducing the cock or cockerel is not advisable, as the introduction of the male at this time will throw them off balance and reduce their egg-production. The best time for the pullets to associate with the male bird is when their combs are beginning to show signs of reddening up for laying, although this does not mean that pullets should be kept unmated until this time. A word of warning — if you intend to run a cockerel with the hens, do make sure that you tell your neighbours and that the local bye-laws permit the keeping of a cockerel.

Those keepers with less than two dozen fowls will have a more relaxed time as there is very little work to do this month. It is the time when all thoughts of hatching and rearing chicks can be put aside until next year. The work of cleaning the house and run should have been completed and all that remains is the feeding of the birds and the enjoyable task of egg-collecting. As long as you realise that the hens must be kept on a balanced layers' mash or pellets regularly, the month of August can be considered as the off season.

September

Those pullets which are due to begin egg-laying at the end of this month should be moved to their winter quarters as soon as possible because they will need time to settle into their permanent surroundings. When the time comes for them to lay, they will already have found a comfortable nesting box in which to deposit their first small egg on the straw or wood chippings provided.

If it is necessary to transfer laying pullets to another house, their egg-laying will inevitably be affected. It can be some weeks before they become accustomed to the change and build up sufficient confidence to start laying again. To avoid pullets becoming stunted in size, it is essential to keep them on growers' food until they are about 6 months old, because this is one of the main factors influencing their development. If this is neglected, then both the yield and the size of the eggs will be affected.

As the evenings are now drawing in, it is a good idea, and a proven one, to feed the corn about 1 hour before dusk. This spreads the hens' feeding activities over as long a period as possible and ensures that they have sufficient time for a good meal before going to roost.

Where the domestic poultry keeper is not following an *ad lib* feeding system, satisfactory progress can still be assured by giving two daily feeds of mash or pellets. This has no adverse effects, nor does it place any strain on the digestive organs. The grain feed given in the afternoon will ensure sufficient nourishment to last the hens through the night and keep the linings of their gizzards in a healthy condition.

All growing birds need a supply of grit and the size of the grit should be regulated according to their size and age. At 4 months old, they should have a trough or box, containing a mixture of ground oyster shell and flint and suitably positioned near their house, so that they can help themselves to these shell-forming materials. The minerals in the grit in no way encourage egg-production but they do provide the elements essential for shell-formation.

If the young birds have been bred from a fine laying strain, it is not unusual for them to lay 2 eggs during 24 hours instead of the 1 expected. As a rule, the second egg is not shelled, but merely has a fine skin covering; there is a risk of this egg being broken, especially if it has been laid from the perch during the night. If this happens, the broken egg must be cleaned up or the hens may develop the habit of egg-eating.

Young birds must be encouraged to spread themselves out on the perches at night, and so plenty of space must be provided. Birds that crowd together, either from habit or lack of room, become overheated and are then susceptible to nasty colds, which will result in loss of condition. Although a good supply of fresh air should be available to the perching youngsters, draughts must be avoided.

Fowls that are well conditioned during the months of August and September will give a good supply of eggs from October onwards. If this foundation period between the summer and the autumn is well used, the hens will produce their eggs throughout the winter, provided that they are given the right food to keep them in good fettle.

October

If September is hot, this will greatly assist the birds, since the heat will improve their blood circulation, thus increasing their energy and promoting a most satisfactory condition in the birds for at least the following 2 months. Once Autumn approaches, it must be the object of everyone who has fowls to see that they are occupied as much as possible; keeping them active keeps the blood circulating well. If their circulation is poor, the fowls, no matter what their age, will become sluggish at this time as the temperatures are slowly dropping. As the weather is beginning to deteriorate, some poultry keepers might like to know whether to close the birds in their houses at night. The answer is 'yes' because closing them in keeps the inside of the house that much warmer. In the mornings, their release will depend on the prevailing weather conditions and the domestic poultry keeper will have to decide whether:

a) to let the birds out in the usual way because it is a normal morning,
b) to keep the fowls in and let them feed from hoppers placed inside their houses because there is frost on the ground or it is foggy or raining. Later in the day they may be let out, unless the ground is covered in snow, for it is good for them to exercise out of doors.

Owners must provide an outlet for the reserves of energy that the birds have and the best way is to scatter some grain on the floor of the house to encourage the hens to scratch about.

November

Now is the time when the poultry keeper who has allowed his young birds to congregate closely together in the corners of the hen house may find that he has to cope with sick birds, suffering from severe colds. When birds are suffering in this way, a bad smell is frequently emitted from the matted feathers under the wings; this is caused by the birds trying to clean away the discharge from their nostrils by placing their heads under their wings. This condition will be aggravated if the fowls are neglected. When the nostrils of the hens are running with mucus, you should clean them by bathing with a solution of permanganate of potash. Make up this solution by dissolving 1 teaspoonful of crystals in 3 tablespoonfuls of warm water. By using a small piece of sponge, or cotton wool buds dipped in a little of the solution, each bird can be carefully cleaned. A clean sponge, or a fresh cotton wool bud, and fresh solution should be used for each bird. To help the birds make a speedy recovery, keep them in the house for about a week. All water troughs should be well scrubbed out daily and, as soon as the birds have recovered, fresh litter should be strewn on the floor.

Apart from the hardship of watching the suffering of the birds, a considerable amount of extra work is involved in caring for them during their illness. It is therefore plain common sense to avoid the problem by paying attention to and eliminating the cause. It is quite a simple matter to prevent this situation by having draught-free houses and giving the fowls plenty of space on their perches.

December

December is the month when the domestic poultry keeper who intends to breed from his existing stock will be collecting eggs, not for consumption but for hatching purposes. As weather conditions can fluctuate considerably at this time of year, it is essential that the birds should be handled each week to ascertain whether they are maintaining a good condition; any loss of weight could indicate an ailing fowl. During the colder periods, do make sure that the birds' drinking water does not freeze; it may be necessary to check this several times a day. It is a fact that hens actually prefer drinking water which is slightly warm.

Introduction

The domestic poultry fowl has existed throughout the world for several centuries, and it is now acknowledged that the poultry we know today evolved from the wild fowl that lived in the primitive jungles and rain forests of the African, Asian and South American continents many hundreds of years ago. The process of evolution in the poultry world has continued throughout the centuries and it is generally recognised that it was during the Roman era that the first of the table fowls were produced through cross-breeding.

Although there are many theories regarding the evolution of domestic fowls, the exact origin of poultry is still unknown. It is, however, assumed that the Jungle Fowl, especially the Red Jungle Fowl (Plate 18), is the most likely ancestor of the domestic birds that we know today. This species is still in existence in the wild and can be found in its natural habitat in the remoter parts of India and Sri Lanka.

Although the Jungle Fowl is widely assumed to be the progenitor of all the domestic poultry that we know today, there still remains some doubt as to whether this assumption is absolutely correct. We know that it was *one* of the forerunners of the poultry fowl, but it can by no means have been the only one. If this has been the case, how would we have been able to produce fowl with not only five toes but also feathering on the legs and feet — essentially a characteristic of Asiatic and Oriental fowl? There are also fowl with different types of combs (Figure 29) and plumage and even some with crests. Because of these differences, poultry today must have evolved from at least four or five different fowl which at one time or another interbred with each other to produce these variations in anatomy.

In the early nineteenth century, great interest was shown in poultry breeding. Many experiments were undertaken during the process of rearing new strains of birds, and these continued for a number of decades. Eventually, for commercial reasons, research was directed towards the selection and development of fowl with a view to increasing meat-production and improving laying abilities.

Figure 29. Although there are several types of combs, the three main variations are (a) rose (b) medium (c) folding.

Over the centuries there have been many people interested in producing fowl solely for their appearance (Figure 30) and colour. Known as poultry fanciers, they have been interested in breeding as near to perfection as possible and to meet the standards required by the Poultry Club of Great Britain. Sadly, however, because of the increase in commercial experiments, the rearing and production of the standard breeds has been somewhat neglected in favour of the 'hybrids', i.e. the results of breeding from close relatives and then cross-breeding with in-bred strains. This cross-breeding has produced several 'hybrid' strains which are better egg-layers or meat-producers, though unfortunately at the expense of colour and beauty. (This is a source of disappointment to the ardent fancier who is mainly interested in rearing pure breeds for exhibition.) Over 40 years have gone into the development of the hybrids and, during this period, the commercial egg-layer has evolved from a relatively inefficient and inconsistent producer into one of the most highly refined products of genetic science existing in the world today.

General research and development, however, has been continued by the dedicated fanciers who, with little or, in some cases, no theoretical knowledge, have produced the many colourful and attractive birds that are available today. However, in recent years, technological advances have finally caught up with advances achieved biologically and, as a result of the ever-increasing costs, good management, the final ingredient necessary for success, is now even more important.

As a result of cross-breeding over the centuries, the new strains of fowl produced vary in size from large breeds to light breeds and bantams, the latter generally being miniatures of the standard breeds

Figure 30. The features of the poultry fowl.

apart from a few exceptions. Because poultry vary in so many ways, in order to have some form of standardisation, a governing body for poultry was formed in Great Britain in 1864, and standards of perfection were drawn up. Several amendments to the administration had to be made, however, and it was not until the year 1877 that the first Poultry Club was formed. This still remains as the governing body for poultry standards.

The basic colours found in the plumage of present-day poultry are black, blue, buff, red, yellow and white. By combining these colours through in-breeding, an enormous variety of patterns can be obtained. There is a beautiful range of colour patterns within the numerous breeds, including the present-day wild birds and, indeed, in the Red Jungle Fowl of today, from which poultry has probably descended.

Breeding

Inevitably, there will come a time when the domestic poultry keeper, especially if there are children in the family, will want to try his hand at hatching and rearing his own chicks. The most natural way of achieving this is to breed hens which 'sit'. Generally speaking, hens can be divided into two categories: sitters (the heavy breeds) and non-sitters (the light breeds). (Bantams are generally considered to be better sitters and better mothers than their larger counterparts.) Poultry can, therefore, be divided as follows:

a) *Sitters*

Asil (Aseel), Antralorp, Barnevelder, Brahma, Cochin, Croad-Langshan, Dorking, Faverolle, Houdan, Jersey Giant, Jubilee Indian Game, Malay, Maran, Marsh Daisy, Modern Game, Modern Langshan, New Hampshire Red, Norfolk Grey, North Holland Blue, Old English Game, Orpington, Plymouth Rock, Rhode Island Red, Scots Dumpy, Sumatra Game, Sussex, Wyandotte and Yokohama.

b) *Non-sitters*

Ancona, Andalusian, Campine, Hamburgh, Lakenvelder, Leghorn, Minorca, Old English Pheasant Fowl, Poland, Scots Grey, Spanish and Welsummer.

Exhibiting

To those readers who are not just interested in keeping domestic poultry to maintain a free supply of fresh eggs for the family but are also considering, or already have an interest in, the exhibition of rare breeds, the following will no doubt be of some interest. For many years, domestic poultry keepers throughout the world have shown a dedication to preserving the rare and pure breeds. These poultry enthusiasts take pride in their achievements by entering and exhibiting their birds in poultry shows. To the newcomer, who intends to enter birds in poultry exhibitions, the scale of points shown will give some guidance as to the standards to aim for, and the relative importance of each feature of the bird. It should be borne in mind that the number of points awarded by judges for a particular feature may vary according to the type of fowl on the show bench.

	Points			Points
Head: Skull	10	Body:	Size	10
Comb	5		Neck	5
Eyes	5		Legs	5
Lobes	5		Feet	5
	25			25
Type: Condition	10			
Plumage	10			
Colour	15			
Markings	10			
Carriage	5	Grand Total = 100		
	50			

For more specific details of points awarded to each individual pure breed, reference should be made to the *Book of British Poultry Standards*.

Chickens

Ancona (Plate 1)

These birds were mainly imported from the Ancona area of Italy in the 1890s because a majority of British poultry breeders had shown great interest in this fowl and were most optimistic as to the progress it would make in Britain. It was at about this time that there was a large demand for a non-sitting hen with a colouring distinct from that of any other hen being bred in the country. However, after several years of breeding, the results were rather disappointing, as no appreciable progress was made in developing the breed because of lack of in-breeding. The Black Leghorn had also made good commercial progress at that time and had become so popular that it was obviously going to be the most profitable of the two breeds. Henceforth, the commercial interest in the Ancona diminished rapidly and it was left to the ardent fanciers to develop the breed further.

The Ancona is a very attractive breed, the main colour pattern consisting of beetle-green and black with a profusion of white tips. The Ancona was found to be particularly wild in habit but also rather shy and really too nervous to keep in an enclosed environment. This contributed to its decline in popularity with the commercial breeder. The Ancona is unmistakably a breed of mottled Leghorn and was produced as a result of crosses between Black and White Leghorns. Nowadays, the mottling of the plumage means that there are difficulties in producing a true colour and this has been one of the drawbacks to the development and progress of this rather attractive breed.

In the cock, the skull is moderately long, deep and of good width. The beak is of medium length with a moderate curve and the eyes are prominent. The single comb should be upright and of medium size, with deep serrations forming five to seven broad-based spikes. The outline should form a rectangular convex curve, the back following the line of the head. In those with a rose comb, the comb should be of medium size, low and square at the front, tapering towards the leader and following the curve of the neck.

The face should be smooth, ear lobes inclined to an almond shape, of medium size (considerably smaller than those of the Leghorn) and free from folds. The wattles are long and fine and the long neck is profusely covered with hackle. The body is moderately long with close, compact plumage, broad shoulders and a narrow saddle. The full rounded breast is carried slightly upwards and the large wings are well tucked up. The full tail should be carried well out. The legs are moderately long with thighs well apart and hidden by the body feathers. The four toes should be rather thin and spread well apart. The bird should carry itself uprightly, with a bold and active appearance. A mature cock will weigh between 2.7 and 3 kg (6 and 7 lb).

The general characteristics of the hen are very similar, except for the comb which falls with a single fold to partly hide one side of the face. The hen weighs approximately 2.3-2.7 kg (5-6 lb). White or cream eggs are laid.

The beak is yellow, shaded with black or, in some cases, creamy white. Preferably, it should not be completely yellow. The eyes should be orange-red with hazel pupils. The comb, face and wattles should be bright red and there should be no trace of white on the face. The ear lobes are white and the legs and feet are yellow mottled with black.

The plumage should be beetle-green with white tips, the more evenly V-tipped with white the better. (It must be *tipped* with white, not laced or splashed.) The under colour is black and all feathers should be black to the roots with a beetle-green surface and tipped with white. There should be no red feathers.

Andalusian (Plate 2)

The Andalusian is classed as a light or medium fowl and takes its name from the southern and most historical region of Spain — Andalusia. Like the true Black Spanish, it is one of the oldest breeds from the Mediterranean area. (Although it is recognised as being of Spanish origin, there are some who believe that the strain was first developed in the British Isles and exported to Spanish territory many hundreds of years ago.) The first of the true Andalusian stock reached the shores of Britain in about the middle of the nineteenth century.

Unfortunately, the breed never really became popular, so little progress has been made in its development since that time and therefore it is still classed as a rare species.

From its make-up, it has the appearance of the old Spanish breed with its almond-shaped white ear lobes and similar expression, but with obvious differences in the colour of the plumage. The cockerel has a very similar stance to that of the Old English Game, with its upright and very alert appearance. Because of its size, the primary function of the Andalusian is laying eggs. It is not necessarily bred for meat-production, although it can reach weights of between 2.3 and 3 kg (5 and 7 lb).

The hen has a single folded comb hanging over one side of the face without covering the eyes, similar to that of the female Leghorn. With a little imagination, one can picture her as a Spanish dancing senorita with dark flowing hair draped serenely over one side of her face and having a proud and confident bearing.

The base colouring of the plumage of the cock is a slate-blue, each feather being distinctly edged with black lacing, apart from those of the neck hackles, which are completely black and quite glossy. The saddle hackles are also black and cover the tips of the primary feathers of each wing. The plumage of the hen has more blue showing than that of the cock and the upper neck hackle is a deep black with a rich lustre effect. The lacing of the lower neck hackle is much more prominent, with a wider black band around each of the hackle feathers.

Appenzeller — or Appenzeller Spitzhauben (Plate 3)

The Appenzeller is one of the oldest breeds of poultry in existence and is reputed to have been first bred by the Vikings and then brought with them when they came to northern France in the tenth century, during the reign of Philippe Auguste. A pure strain was subsequently developed in or near the town of Appenzelle, which is situated in the lower regions of Switzerland, near to the German and Austrian borders and not far from the upper reaches of the River Rhine.

The Appenzeller is one of the newest imports into the British Isles and is still considered a rare species. Fortunately, there are a few poultry fanciers who are devoting their efforts to developing new strains of the breed. During the war years, the breed nearly died out but, fortunately, in the late 1940s and early 1950s, a number of local farmers in Appenzelle again began to breed the hens and consequently there has been a very gradual progression in their development. Nevertheless, it is still considered to be quite rare. As a result of this situation, a specialist club was formed, under the auspices of the Poultry Club of Great Britain, to further the interests of the breed.

There is a legend that tells of an old crofter, who lived in the wilds of the Bernese Oberland. He possessed a small piece of land on which he grew vegetables. He also kept goats to provide him with a free supply of milk and some hens to supply him with eggs and meat. He was a contented recluse and was able to supply himself with all the necessary vitamins and minerals needed to sustain him. It is thought that the breed of hens which he reared was the Appenzeller Spitzhauben (meaning 'laced-Appenzeller' or 'Appenzeller laced-bonnet'). The name is undoubtedly appropriate, as the city of Appenzelle is noted for its lace.

Since that time, and as a result of the dedication shown by the fanciers of the breed, other strains have now been produced, e.g the Silver-Spangled, Gold-Spangled, Black and also the Barthuhner.

The Appenzeller is a very strong fowl and quite hardy because of the high altitude conditions under which it was first bred; consequently, it does not mind the colder climates.

Weights vary between 1.4 and 1.8 kg (3 and 4 lb) for the cock. The hen has an average weight of about 1.4 kg (3 lb).

Araucana (Plate 4)

This is a very ancient fowl, dating back far beyond the Elizabethan period. During the sixteenth century, when the Spanish ships were roaming the high seas searching for new lands, they took with them several of their own breeds of poultry to supply them with a regular diet of fresh food in the form of eggs and poultry meat. They also intended to introduce the fowls into the new lands which had already been discovered. On their arrival in Chile, however, they discovered that they were not the only people to keep domesticated fowl and, during their visit to the Arauca region of northern Chile, they noticed that the Araucana Indians had their own particular breed of poultry, which laid pale bluish green eggs. This was the Araucana. The Araucana Indians soon cleared the Spaniards, and their poultry, off their territory, so that today the Araucana fowl still remains the true breed of Chile.

The crews of the Chilean merchant ships of yesteryear also kept Araucanas on board because of their prolific egg-laying qualities. In some instances, hens have been known to lay 280 eggs in one year, which is above the average even for the commercial egg-layer of today. These merchant seamen manned the ships which supplied the different nations with guano, which was regularly collected from the small islands near the Chilean coast. At one time in its history, the Araucana was thought to be a Norwegian fowl because it was discovered aboard a Norwegian sailing ship that had foundered off the northern coast of Scotland. This theory, however, was soon discounted.

As the maritime nations began to trade with South America, the Araucana found its way into other countries and, inevitably, cross-breeding developed new strains. It was in the late 1920s that George Malcolm created the true Lavender Araucana, but it was not until the early 1930s that it first became standardised by the Poultry Club of Great Britain.

The Araucana is classed as a light breed of poultry that can produce weights of approximately 2.7 kg (6 lb) for the cock and 2.2 kg (5 lb) for the hen. The strain of Araucana in existence today can be said to be man-made. It is the Rumpless Araucana that is the true and original breed of Chile.

Asil or Aseel (Plate 5)

The Asil is one of the oldest breeds of poultry from the Far East, with an ancestry dating back more than 2000 years. They were greatly admired by the Maharajahs of India in days gone by because, like the Old English Game, they too were fighting birds and were widely used as a source of entertainment. Although the barbaric sport of cock-fighting was outlawed in Great Britain in 1849, it still takes place illegally. Of all the game breeds, it is the Asil which is the most powerful and dominant, so much so that, if all breeds were placed in one pen, the Asil cocks would be the only birds remaining, as they always fight to the death. It is the aristocrat of all fighting cocks. Indeed, in the Asian language, the word *asil* means 'aristocrat' and it is certainly worthy of the title.

Although it is a very popular breed in India and Sri Lanka, it is not solely a native of those lands, as similar types of fowl exist all over the

Asian continent. For many years, it was used by the people of India and neighbouring countries as a domestic bird, not for the eggs, as they are very poor layers, but for the abundant supply of breast meat. Being a good meat-producer, it has been widely used for crossing with hens that have a very large carcass, such as the Sussex and Rhode Island Red, which also have a large and particularly wide breast.

It is thought that the Asil was imported into Britain about 200 years ago. King Charles II became very interested in this new fowl and arranged the import of some of this very rare breed for himself; these he kept at his Court. The introduction was not easy and several of the newly acquired flock were lost because they did not acclimatise well to English temperatures, which are appreciably lower than those of their native land.

When kept apart from cocks of other breeds, the Asil is a docile and charming fowl by any standards and can quite easily be made into the safest of children's pets. It is very intelligent, possibly one of the most intelligent of all breeds, and very easy to handle under normal conditions. The Asil hen makes the most wonderful broody but, alas, lays very few eggs; sometimes a hen may lay only 2 or 3 eggs in a year and can incubate only 7 or 8 eggs in a sitting. Because of this it will never become a commerical fowl.

There is some controversy as to the correct spelling of the name of the breed; sometimes it is spelt Asil and sometimes Aseel. Asil is perhaps the nearest to being correct but, in Britain, it is sometimes spelt with a double 'e' to indicate the correct pronunciation.

Asils, when bred in India, can reach weights of about 6.8 kg (15 lb), but for those bred in Britain the standard weight is usually only between 2.2 and 2.7 kg (5 and 6 lb).

The cock has a very erect attitude and stands firmly on both legs, which are naturally positioned wide apart. A very muscular and powerful bird, it is always alert and moves quickly. The head and neck are pushed slightly forward from the body, as if in an attacking position. The neck is very powerful and prominantly curved near the top. The bright staring eyes are always on the alert, looking for any possible foe. The beak is short and very effective in battle because of the powerful jaw bones.

The legs are reasonably long for the size of the bird but the thighs are moderately short and very muscular. The four toes are straight, with curved toenails that, together with its natural spurs, can be lethal in attack. The broad back tapers towards the tail and is well developed and powerful. The wings are positioned slightly outwards near the shoulders which are also very muscular. The plumage is rather short and hard to the touch — by no means fluffy and soft like other breeds. The unusual feature of the Asil is that it has patches of what appear to be raw skin showing on the wing joints and thighs. The main colour of the plumage is dark red, although some variations have been produced.

Apart from the normal sex differences, the hen is very similar to the cock in general appearance.

It should be noted that, when breeding Asils, they will invariably fight with each other. Even the young chicks, after the age of 4 weeks, will vigorously display this natural aggressive instinct. Therefore, once they have become independent, they should be placed in separate pens. Although it is a fighting bird, an Asil cock can be run with another flock as long as no other cockerel is included.

The Asil crow is quite a short one and so it is possible to keep these poultry in towns and cities, without fear of annoying the neighbours..

Barnevelder (Plate 6)

The Barnevelder is a heavy breed of fowl which was introduced into the British Isles in the early 1920s; by the latter part of that decade, it was beginning to increase in popularity with the breeders because of the brown eggs which it laid. This bird came from Holland and was first bred in the district of Barneveld, hence its name. The Barnevelder was produced as a result of cross-breeding the Croad-Langshan with the Golden-Laced Wyandotte and, in the early stages of breeding, two specific varieties evolved: the Partridge and the Double-Laced. A Black variety was also produced at that time, but it very quickly diminished in popularity as the earlier breeds were so attractive.

The most popular features of the Double-Laced are its bold upright stance and its compact and richly shaded plumage. The body is of medium length with broad shoulders and wings well tucked in. The breast is full and rounded and the tail has a graceful sweep like the hands of a ballerina. It is generally recognised that Barnevelders are fast-growers and, because of this, the cockerels are acceptable for the table after about 18 weeks. The Barnevelder pullets are some of the best layers that can be obtained from the true breeds of poultry and they have also proved to be very good sitters. In fact, they are better than most of the other true breeds of fowl, with perhaps the exception of the Rhode Island Red.

One special feature of the plumage of the Double-Laced variety is the greenish black shading, with some ground colour, and the partridge-pencilling in the hens and pullets also has a most striking appearance. Its boldness and agility, as well as alertness, add to its popularity.

The difference in the markings of the Double-Laced and the Partridge is in the blending of the two shades of colouring. The feathering of the Double-Laced hen has two sets of lacing, i.e. greenish black at the edge, a surrounding area of reddish golden-brown, then a second lacing of greenish black. Finally, the centre of the feather is also reddish brown. In the Partridge variety, the centre of each feather is a reddish golden-brown inside the lacing of greenish black.

Brahma (Plate 7)

It is somewhat difficult to authenticate the history of this breed, as it is impossible to substantiate any of the evidence. In earlier poultry writings, it was believed that the Brahma fowl was the result of crossing the American strain of the Chittagong fowl with the Cochin; however, there is little evidence to support such a claim. Nevertheless, it is conceivable, because the Chittagong and Cochin breeds share some characteristics with the Brahma. Even when the first of the Brahma strain was brought to the shores of Britain from the USA, during the reign of Queen Victoria, confusion as to its authenticity still prevailed and it was eventually accepted that this American breed was, in fact, closely related to the Chittagong and was not the true Brahma.

Further evidence to indicate that the true Brahma was from another part of the globe is given in the *Illustrated Book of Poultry*, which states that a mechanic by the name of Chamberlain first brought them to Britain. He apparently became acquainted with a sailor, who had told him of the large hens that were kept by families in the States of New York and Connecticut. Chamberlain was so impressed with the sailor's description of the hens, that he gave him money so that, on his return to the USA, he could purchase a cock and two hens on his behalf. In May 1847, the sailor bought three pairs of the breed and, during conversation, the vendor made some reference to the fact that these large fowl had first been brought to New York by the sailors who manned the 'India' ships that traded between the two continents.

The sailor soon returned to England and delivered the new fowl to their owner. Mr Chamberlain on seeing them noticed that each pair had a different coloured plumage (light grey, dark grey and red) and not long afterwards began to realise that these new fowl were similar, if not the same, as the American Chittagong fowl. He therefore concluded that his new flock were the result of crossing the Chittagong, also of Indian descent, with the original and true Brahma, which had been brought to the United States by the merchantmen that sailed the high seas in the late eighteenth century.

Because of its characteristics, the true Brahma, with its leg feathering, is undoubtedly of Asian origin and, during the earlier days of the breed's development in Britain, it became known as the Brahma-Pootra (Brahmaputra) — the name of the river in East Pakistan, which flows through the area near to the borders with Assam, where the original strain was first bred.

In the late nineteenth century it became extremely popular and, because of its high standing in the world of poultry at that time, it was given the honorary title of 'Lord of Creation'. As time passed, its popularity started to wane with poultry keepers and so it never really became a major commercial venture. Its popularity continued to decline for many decades until the dedicated fanciers decided to take a hand in ensuring the survival of the species and so today it is bred purely for exhibition.

The Brahma, like the Cochin, is a heavy breed of Asian descent, that lays tinted coloured eggs. Because of the rarity of the Brahma, the plumage has few variations. The true colours are the Buff/Columbian, Dark, Golden, Light and the White. Classed as a heavy breed, the weight of the cock can be as much as 4.5 kg (10 lb) and sometimes even more, whilst the hen weighs in the region of 2.7-3.6 kg (6-8 lb). Therefore, it could well have been a suitable bird for the table.

Campine — Silver (Plate 8)

The Campine is one of the ancient breeds of Belgium and is a favourite with the ardent fancier because of its plumage and delightful appearance. It is not, however, accepted as a general domestic fowl because of its poor egg yield, which is the result of several years of in-breeding to produce the attractive plumage. As the breed is widely exhibited in most of the major poultry shows throughout the British

Isles, the following will undoubtedly be of interest to the fancier.

The cock has a graceful bearing but is always on the alert. The skull should be moderately long, deep and reasonably broad. The eyes are rather prominent and the beak is short. The face is smooth, with ear lobes of medium size, free from wrinkles and inclined to an almond shape. The wattles are long and fine. The neck, which is well covered with pure white hackle feathers, is of a good length. The body is broad with a full rounded breast and the long back narrows towards a well spread tail, which is proudly carried well out. The large wings are neatly tucked in. The legs are reasonably long and free from feathers and the slender toes are well spread apart. The average weight is approximately 2.7 kg (6 lb).

In general, the hen is similar to the cock, apart from the comb, which falls over the hen's face, as it does in the Leghorn, giving the bird a rather sophisticated appearance and adding character to the face. The eggs are white. The average weight for the hen is 2.2 kg (5 lb).

The beak is horn-coloured and the eyes are dark brown with a black pupil. The comb, face and wattles are bright red. Legs and feet have a bluish tinge and the toe nails are the same colour as the beak. The plumage of both the cock and the hen is pure white over the head and neck hackle. Elsewhere the plumage is also white but this acts as a background for a strong beetle-green barring.

There is also a Golden variety in which the head and neck hackle are a rich golden shade. This, like the white of the Silver Campine, acts as a golden background to the beetle-green barring.

Cochin (Plate 9)

The Cochin is a very ancient breed of fowl and was originally imported from China almost 200 years ago. At that time, the breed was regarded as a most unusual species because of its feathered legs. However, as it became more widely known, it was soon realised that it was a breed with good laying habits, producing brown-shelled eggs. Apart from these excellent laying qualities, the Cochin is also a heavy breed of fowl and at one time was recognised as a very acceptable table bird.

As a result of several decades of cross-breeding, there are now numerous strains in existence, e.g. the Black, Blue, Buff, Cuckoo,

Partridge, and the White. The first breed to be introduced into the British Isles was the Buff Cochin, with its buff- or cinnamon-coloured body plumage and a black-tinted neck and reddish hackles and saddle. The plumage of the cockerel is of several blended colours but still retains the red hackles and saddle.

As time passed and their laying qualities became more widely known, fewer people cared to breed them just for the quality and beauty of their feathering and, as a result of the cockerel being allowed to run with the general farmyard hens, several other half-breeds of Cochin emerged. It was soon discovered that these half-breed Cochins started laying at a much earlier age than the pure breed. It was also realised that their rate of growth was much faster and they could, therefore, be ready for the table much sooner.

The Buff Cochin was, in fact, the forerunner to the Buff Orpington. It occurs in several shades of buff and, for exhibition purposes, the colouring should be free from blotches or discolouration.

The Partridge Cochin has the typical colouring of the ancestors of the domestic fowl. The male has a neck and saddle which is a rich bright red colour with a dense black stripe along each feather. The back, shoulder and wings also have a rich red colour but darker than that of the neck and saddle and with a wide greenish black stripe across each wing. The end of every feather is tinged with black.

Dorking (Plate 10)

The Dorking fowl is one of the oldest breeds of the British Isles and its history extends back as far as Roman times. Over the centuries, it has played a prominent part in cross-breeding and was regularly used in the development of new strains. Nevertheless, the Old English Dorking never really became popular in relation to the other new breeds developed, because of its short legs and fifth toe. As the birds were developed through breeding, the fifth toe eventually became rather long, which proved to be a disadvantage in walking; as a result, the popularity of the breed started to decline. Because of this decline, few new strains were produced and, unfortunately, the bird became in-bred. This in-breeding weakened the egg-producing organs which, in turn, led to reduced egg-laying qualities. This problem invariably occurs with all in-bred strains, which tend to be of the less popular varieties.

The Silver Grey Dorking is recognised as holding the premier position of being the oldest English breed of poultry. Although it is a good layer, it can still be classed as one of the better table birds of the true breeds. The Red Dorking, also, has a most eyecatching appearance because of its colourful attractive plumage.

The Dorking breed has a quiet temperament with a stately and proud appearance. It has a rather long and deep breast, which is well formed. This breed should not be reared on a hard standing area, such as wooden floors or concrete, since the position of the fifth toe creates some difficulty for the birds when standing on flat surfaces, which eventually weakens their legs. It is best to rear them on soft earth or to have plenty of straw or scratching material on the floor. Deep litter is certainly advantageous.

In the cock, the head is large and rather broad with a short but stout beak, well proportioned with a slight curve. The eyes are wide and full. The well bred domestic bird has a single well defined comb, especially the Red or Silver Greys. The comb should be very erect, large and evenly serrated and the rear end should be slightly upturned. The face should be quite smooth. The ear lobes are reasonably well developed and about one-third the length of the wattles, which are quite long and pendant in shape.

The neck is rather short and covered with an abundance of hackle feathers, which fall well over the back and give the neck a very broad appearance. The body is quite large, long and deep, and is rectangular in appearance The feathers are tucked in rather tightly. The breast is well rounded and rather full. The wings are large and well tucked into the sides. The full sweeping tail flows with character, just like the sweeping hands of a ballerina, and is carried well out. The Dorking should have a very proud and stately appearance with the breast pushed well forward. A good well-bred cock can weigh anything between 5.4 and 6.4 kg (12 and 14 lb).

The hen is very similar to the cock, apart from the normal sex differences, and weighs between 3.6 and 4.5 kg (8 and 10 lb).

The beak can be white or horn-coloured. The eyes are bright and the comb, face and wattles are red. The legs and feet of the true breed, including the nails, can be almost pure white.

The Silver Grey Dorking is recognised as the leader in popularity of this breed. In the cock, the hackle (i.e. the neck and saddle) can be silver-white but should be completely free from light brown or rust or similar markings of any kind; a strip of silver grey in the lower feathers may be permissible but not necessarily desirable. The hackle and back should be silver-white and free from any yellow striping or tinge. The upper part of the wings is silver-white and the lower half is black with a blending of green tinge or alternatively a bluish gloss. The outer secondary feathers are pure white and inner web is black; there should also be a black spot at the end of each feather. The primary feathers are a rich black with a white edge. The remainder of the plumage is a rich black and completely free from any form of mottling.

In the hen, the hackle is silver-white with the lower feathers tipped with black. The breast colouring is salmon-red, gradually fading to light grey on the upper leg. The wings and remainder of the body are silver-grey, finely pencilled with a darker shade of grey. The tail is a much darker grey, the outer feathers being charcoal grey or even black with a white edge. The remainder of the plumage is a rich black and completely free from any form of mottling. Other strains in the Dorking breed are the Cuckoo, the Dark, the Red, and the White.

Faverolle (Plate 11)

The Faverolle is a most attractive breed of poultry, which is of French origin and was first bred in the village of Faverolles in northern France, hence its name. They are recognised as being good layers and are excellent table birds. However, they are not noted as a popular breed, mainly because of their feathered legs — during wet days on muddy ground, they get splattered with mud spots and look a sorry sight. The breed was first brought to Britain during the late nineteenth century and, at that time, was bred mainly for the table. However, because of their leg feathering and rather long body, which lacks symmetry, the breed rapidly declined in popularity. Nowadays the Faverolle is bred mainly for the show ring. The leg feathering is the result of cross-breeding the Dorking, Cochin and Houdan with the Brahma.

It has an active and alert appearance but also a rather shy expression. Its body is elongated with a deep keel, very broad shoulders and flattish back. The breast is very full and well rounded. For the size of the bird, the wings appear to be quite small, probably because they are carried closely tucked in to the body. The tail feathers are quite long with an upward sweep, giving a flowing effect.

The beak is rather short but slightly broad. The single comb is of medium size and upright with a fine serration and very smooth. The wattles and ear lobes are quite short and are invariably hidden by the muffle or beard. The legs are rather short but stout with thick thighs set well apart. The breed has five toes, but unlike that of the Dorking the fifth toe is quite short and curls slightly upwards.

The hen differs from the cock in two main respects: the neck is much more upright than that of the male and the tail is carried at an angle of about 45° to the line of the back. In all other respects, the sexes are very similar.

The main scoring points of the breed when entering in shows and competitions are related to the bird's size and condition, type, colour, beard and muffling. These features usually have a maximum scoring rate of approximately 20 marks each. The comb, foot-feathering and feet can each score a maximum of 5 points.

The cock of the Salmon variety is a most attractive bird with a black beard and hackle and neck feathers of a light straw colour. Its shoulders, back and wing tones are a rich mahogany, whereas the breast, thighs, tail and leg-feathering are a rich black. The wing bar and secondary feathers are black and the primaries have a white outer edging with black tips. The eyes are a hazel shade and are quite prominent.

There are several other strains of the Faverolle. The Black has black feathers with a rich beetle-green sheen in both sexes and a black beak. In the Blue, the cock has a black tail and dark blue wing bows, while the remainder of the feathering is a rich blue, laced with a slightly darker shade. The hen is a uniform rich blue with lacing as in the cock but invariably with legs and feet also with a bluish tint. In the Buff, the plumage of both sexes consists of a rich lemon-buff feathering over the entire body.

Other strains are the Ermine and the White. The White has pure white feathering over the entire body with a white or off-white beak, hazel eyes, red face, partly muffled, and white legs and feet.

Frizzle (Plate 12)

Although the breed is thought by many to have its origins in Japan, there is some speculation as to the truth of this. There are many poultry fanciers who also believe that the Frizzle, with its quaint curling plumage, was in fact produced accidentally as a result of breeding and cross-breeding experiments in years past. Another school of thought says that the Frizzle was first imported from the West Indies and the Caribbean regions. The only certain thing is that the exact origin of the Frizzle still remains unknown.

It is a unique fowl insofar as its feathers appear to lie the opposite direction to that of any other breed, i.e they curl backwards and point towards the head. Despite its strange appearance, the hen is a good layer and a good mother. Unfortunately, as a large fowl, the breed has almost ceased to exist in Britain; nevertheless, there is still a plentiful supply of the bantam kind. The Frizzle is by no means a general domestic or utility fowl and nowadays is bred purely for exhibition.

Since the introduction of the pure White, other varieties have gradually been brought into existence, e.g. the Black, Blue and Buff, as well as the Columbian strain, the Duckwing, Black-Red, Brown-Red, Cuckoo, Pile and Spangled. Of all these different colour varieties, it is the White and the Black which are still the most common. The colour of the plumage is the same in both cock and hen, which is not the case with many other breeds.

Apart from the usual sex differences the cock and hen are similar. The body is quite broad for the size of fowl, with a full and well rounded breast. Other parts of the anatomy are in equal proportion in relation to the size of the fowl. The eggs are sometimes pure white, sometimes tinted. The main attraction of the bird is its abundance of frizzled plumage.

Game — Indian and Jubilee Indian (Plate 13)

Some controversy still prevails over the exact origin of the Indian Game. Undoubtedly, the breed that we know today did originate in England and its history dates back to the days when the Cornish tin mines were a hive of industry and many of the Cornish miners were poultry fanciers. Although the origin of the bird is British, it is the result of crossing the Malay (an Asian fowl) with the Old English Game and mating the offspring with the Asil; the result was a new strain of Indian Game. The Asil and Malay are, of course, the true Indian Game, being natives of India, Sri Lanka and other Indonesian countries.

It is also possible that the Indian Game may have come to Britain during the nineteenth century, when India was part of the British Empire. It is conceivable that the British took with them to India a number of Old English Game birds, not only for entertainment but also because they were considered to be good meat-producers. Other strains may well have developed as a result of the Old English Game breeding with Asian wild fowl and eventually some of these new strains may have found their way, illegally, into Britain. There is a tale, which dates back to the late nineteenth century, about a soldier returning from India, who stopped over at a friend's house in the Lebanon and was shown around the poultry yard because of his interest in domestic fowl. The unusual breed that met his eye appealed to him and it is thought that, eventually, some of its offspring were imported into England from the Lebanon.

Nevertheless, much of the credit for the development of the Indian Game must go to the dedicated fanciers of Cornwall — the county recognised as the birthplace of the Indian Game that we know today.

Although the Indian Game is by no means an ancient fowl, its ancestral line does extend back for 2000 years or more. The breed which is recognised by the Poultry Club of Great Britain has in its make-up about 25% Old English Game, 25% Malay strain and 50% Asil. Not surprisingly, there is a strong infusion of Asian blood.

Even today, in the eyes of the rare-breed enthusiasts, this game fowl still enjoys a measure of popularity as an excellent table bird because of its great quantity of breast meat, and a fair number are now kept for pure-breeding. There is still a reasonable demand for the cockerel for cross-breeding, in order to produce good table fowls. However, a fair price has to be paid to purchase a strong healthy specimen because the number produced is still rather small. Those who already have experience in poultry breeding will know of the popularity of crossing Indian Game with the Dorking breed to produce an excellent table bird. Such birds were bred for many years for the best tables in the land.

When considering the general results of this cross, it should be appreciated that the offspring were never recognised as very good layers and their only recommendation as utility birds was their good meat-production. To enable the fancier to continue to breed and improve the strains, obviously, eggs must be forthcoming. As it is not natural for the Indian Game and Dorking to lay during the winter months, strict selection must be made and only the best layers should be used for breeding.

Indian Game cocks are very active fowl and are generally recognised as especially good stock birds. A well-formed male, in good condition, will fertilise 12 or more hens on a free range, or about 8 or 10 if reared in a confined space. The hens are known to be marvellous sitters and also good mothers and they will protect their chicks from anything that might attack them — cats, vermin or the like, even human beings.

Like the Asil and Malays, the Indian Game has a very upright stance, powerful and broad shoulders and a very large breast. It is a very active fowl and very courageous. The head is held high, giving a powerful and confident appearance, and is broad, with a well curved beak. The eyes are full and constantly alert. The comb is of the pea type. The neck is short to medium in length and arches slightly near the head but not in quite so pronounced a fashion as in the Asil. The legs are thick and very strong, well set apart but not as long as the true Asian breed. The feet are strong and powerful with long toes. The feathering is rather short, firm and hard to the touch, fitting tightly to the body.

In the cock, the main plumage is black with a beetle-green sheen with a lustre of glossy effect and tinges of chestnut colouring in the neck hackle, wing and tail feathers.

The main colouring in the hen is chestnut-brown with black or beetle-green in the neck hackle and throat plumage. There is double-lacing on the breast, underbreast and thighs, with a general black or beetle-green lacing over the remainder of the plumage.

Jubilee Indian Game
It was in the late 1890s that the Jubilee Indian Game was first developed by Henry Hunt who hailed from Gloucestershire. Hunt was a dedicated geneticist and naturalist and it was he who produced the Jubilee strain to commemorate the Diamond Jubilee of Queen Victoria — hence its name. In the early 1900s, the breed became very popular in the USA because of its quick development and good meat-producing qualities, which made it an ideal table bird. In addition, it soon became recognised by the fanciers as a fowl worthy of exhibition. As the import of the breed to the USA increased dramatically, due to its popularity, it was eventually renamed after its place of origin and is now known there as the Cornish Fowl.

It is classed as a heavy fowl because of its tough and broad appearance. Its general characteristics are similar to those of the Indian Game. Fully mature birds can reach weights of 2.9-3.9 kg (6½-8½ lb).

The cock is white with chestnut markings on the hackle feathers, wing bars and primaries and a touch of chestnut or bay markings on the saddle hackle.

The hen has a ground colour of mahogany or chestnut, the throat and hackle feathers being white. The secondaries have similar markings. The main tail feathering is white throughout.

The fowl which are recognised as Modern Game are generally kept by the dedicated fancier, who endeavours to achieve perfection in colour. There are such a large number of Game varieties that it is not possible to describe all of them in this book; however, for the benefit of readers who are interested in this breed of poultry, the most popular of the Modern Game varieties are described below. Other varieties include the Birchin, Pile and the Brown-Red fowls.

The Modern Game is classed as a heavy breed and lays tinted eggs. It has evolved as a result of the dedicated fanciers cross-breeding with the Malay strains.

Black-Breasted Red

This is the most common of all the varieties. The cock should have a black breast and tail, and the lower part of the body should also be black. The hackle and saddle feathers should be of a rich brownish red or orange colour, whereas the upper section of the wings should be of a deep red with black edging, the centre section being black and, at the end, a reddish brown shade very similar to the colouring of the hackle feathers.

The hen has a reddish or light partridge-brown back and wings, and the underpart of the body is a lighter shade.

The head of the cock should be orange-red, the beak dark green and the eyes, comb, face, wattles and ear lobes a bright red. The legs should be greyish black or willow shade.

Golden Duckwing

The cock bird should be black-breasted with a creamy white hackle and golden-orange wing bows. The wing ends or bays should be white. Both sexes have a dark horn-coloured beak, with bright red comb, face, wattles and ear lobes, and eyes of a rich ruby colour.

The hen, however, should be a delicate steel grey in colour with a salmon breast and white hackle, lightly striped or pencilled with black.

Silver Duckwing

This variety of Modern Game is similar to the Golden Duckwing except that the cock has a silver white hackle and shoulders and the hen has a lighter shade of grey on the body.

History dictates that the Old English Game Fowl was in existence long before the Romans invaded Britain in A.D.43. This magnificent breed is generally recognised as being of true British origin, and the Black-Breasted Red is undoubtedly one of the most beautiful varieties with its profusion of colours.

It is written in the annals of Roman history that the English kept fighting cocks for pleasure and entertainment, rather than for domestic purposes. To a degree they were right, as cock-fighting is known the world over and, even today, still takes place, although to a much lesser degree. In 1849, Parliament decreed cock-fighting to be illegal throughout Britain. Nevertheless, it is still practised, albeit secretly, mainly by the gypsies and travellers roaming the English countryside.

The strains of the Old English Game breed are too numerous to mention in detail in this book. For instance there is the Black-Breasted Black-Red, Black-Breasted Red, Black-Breasted Silver, Black-Breasted Yellow, Black-Breasted Brown-Red, the Pile, Spangled, White, Cuckoo and Brown-Breasted Yellow, to name but a few. Of all the strains, the Black-Breasted Red is considered to be the closest in colouring and character to its ancestor, the Red Jungle Fowl.

In the cock, the breast is full and rather broad with prominent muscles on either side. The wings are long in relation to the size of the body and quite powerful, the bird being quite strong in flight. The feathers of the large tail are carried well upwards and separated. The small hawk-like head has a tapering beak, slightly curved and very strong. The large fiery and fearless eyes are always very alert. There is a single upright comb and the ear lobes and wattles are quite fine and thin in relation to other breeds. There is indeed a marked difference in its make-up to that of the Modern Game. The neck is long and very strong and powerful.

Obviously in the Old English Game fighting cocks the legs are very strong, as these are the weapons for attack and defence. The thighs are rather short with powerful muscles. Shanks are also long and very strong. The four toes are reasonably thin and tapering, terminating in long curved nails which are set firmly on the ground and spread slightly apart.

One of the interesting habits or mating actions of the Old English Game cock is that, when he is feeling lonely or requires the attention of the females, he will employ his courtship routine, which is to move his head up and down quickly in an exaggerated pecking action with a simultaneous clucking sound. This mating call attracts the attention of the hens and, when they see the cock pecking for food, or simulating such an action, they hurriedly rush over to see what they are missing.

A word of warning: when running an Old English Game cock with a flock of hens, under no circumstances should you enter the pen and handle any of the hens while the cock is running free. His instinct is to defend and, if such a situation occurs, as it once did with the author and his wife, the cockerel will immediately fly at the intruder with his claws held up in the attacking position. If it is necessary to handle the hens for any particular reason, then the cock or cockerel should first of all be isolated in the hen house or removed from the pen completely. This situation, of course, does not necessarily apply to other breeds of poultry, but even so care should always be taken.

The hen is very similar to the cock, except for the tail, which is slightly fan-shaped. The eggs are tinted.

The plumage on the thighs and keel of the cock should be black. The tail is also black, but with a beetle-green sheen in some of the tail feathers. The hackle and saddle feathers are bright orange-red and the shoulders and wing bow a rich dark red, almost claret. In the Oxford type, the legs have a red stripe from the knee to the ankle. This stripe is not noticeable in the recognised Carlisle type.

Hamburgh (Plate 16)

The history of this breed is almost non-existent and the place of origin is still undefined. One thing, however, is certain: it did not originate from the town in Germany whose name it bears. It has been on the decline for decades, which, in a way, is rather surprising, as it has always been recognised as a prolific layer. It has not, however, been classed as a very good sitter and its non-sitting qualities, together with the rather small size of the eggs, may have contributed to its lack of popularity. Although known as a non-sitter, occasionally a Hamburgh hen will become broody, as sometimes happens with commercial strains.

When the Leghorn was introduced into Britain in 1872, it was immediately successful and quite overshadowed the Hamburgh. This situation has continued to the present day, partly because the majority of breeders do not like a rose comb and also because the Leghorn lays much larger eggs.

Hamburghs are a very robust breed that live much longer and lay up to a greater age than any other known breed. Ten years has been recognised as the average laying period and even longer periods have been recorded for the pure breed. A great many of the general breeds of hens today seem to end their laying careers after their fourth or fifth year.

The varieties of the Hamburgh breed which exist today are the Gold-Pencilled, Silver, Gold-Spangled and Silver-Spangled. All of these have the unusual rose comb, which is attractive and a point of beauty, which makes it stand out from other breeds.

Houdan (Plate 17)

The Houdan is a fowl of unusual appearance, with a unique head-dress and beard, and was named after the French village of Houdan. It is recognised as being one of the oldest of the French breeds and was introduced into the British Isles during the mid-nineteenth century, preceding the import of the Faverolle by some 30 years.

Because of its head-dress and beard, the Houdan can become rather unsightly, particularly during inclement weather. If the bird is allowed to scratch for food on wet and muddy ground, the crest feathers become wet and tend to droop over the front of the head. This often causes irritation and, if the bird then scratches its face, inflammation and infection may result. Swelling of the face may follow and, if this is not dealt with immediately, may become a major problem. Because of the wet feathers covering the head, such problems are not readily noticed and it is advisable to keep the birds in dry conditions at all times.

The Houdan was originally developed as a table bird and is classified, for general purposes, as a heavy breed. The cock and hen weigh, on average, about 3 kg (7 lb). The hens are good layers but are not generally recognised as good sitters, which is probably the reason for the breed's decline in popularity.

For exhibition purposes, although the large crest and fifth toe can be somewhat troublesome features, it is necessary to breed with these two features in mind.

The plumage of both cock and hen is a glossy beetle-green and black, flecked or mottled with pure white spots, which should be evenly distributed. The legs and feet are off-white, mottled with black or blue markings.

La Flèche is a very old breed that has its origins deep in France. Because of the lack of interest shown by the geneticists and poultry fanciers over the past decades, its development has been somewhat retarded, so much so that it has been classified as a very rare breed for very many years. In spite of its devil-like appearance, with its large horned comb, which gives it an air of fearsome power, it never really created much interest with the poultry enthusiasts of the British Isles.

More recently, the British and German geneticists and fanciers, like those in the USA, have renewed their experiments to improve the colour and appearance of the breed but, alas, without much success. The Black is still the predominant variety in Britain, although, in Europe and the USA, the Blue and the White are also found; these colours, however, are very seldom seen in Britain. As the European poultry fanciers tend to place more emphasis in experiments for colour rather than quality, other strains will doubtlessly emerge.

At some stage in its history, the breed must have interbred with the Poland or Polish, because of its horn-like comb. In spite of its lack of crest, it is also related to the all-black Crève-Coeur, another old French breed, which, in turn, has some resemblance to the Houdan.

In the 1840s, in France, a country noted for its fine delicacies and haute cuisine, together with its love of poultry meat, experiments were encouraged with the hope of producing a fowl that would have an abundance of succulent white breast meat suitable for the Paris market. La Flèche figured strongly in these experiments, with a reasonable amount of success.

La Flèche is a fowl which has a more upright stance than the standard attitude, with its large and prominent breast well pushed out. The back, which is reasonably wide, slopes at an angle of about 20° from the shoulders. The cock has large and well curved sickle feathers that are raised quite high in relation to the body.

The most unusual feature of La Flèche is its devil-like comb, consisting of two spikes standing almost upright, and its large and deep nostrils. Its ear lobes are pure white. The colour of its beak, eyes, legs and feet is almost slate black. Although it is related to the Crève-Coeur and the Polish, La Flèche is completely free of any crest.

The breed is considered to be a large fowl, reaching weights of between 3.6 and 4 kg (8 and 9 lb) for the cock and between 2.5 and 3 kg (5½ and 7 lb) for the hen, when fully developed. The hen lays fairly large white or tinted eggs.

In the British Isles, only the Black has a standard and the colour should be a deep rich black with beetle-green sheen. The Blue variety has a slate-blue shading and the White has entirely white plumage. These two varieties have yet to appear in Britain.

Lakenfelder — or Lakenvelder (Plate 20)

The history of the Lakenvelder is a little obscure to say the least, and it can only be classed as a European fowl which seems to have originated in an area near to the Dutch/German borders. It is most probably more Dutch than German, as the word *lakenvelder* means 'field of linen' in the Dutch language. It is a fairly old European breed, developed from a selection of the light-weight European fowls that were inter-bred centuries ago, probably by accident. These were mainly the horizontal breeds that had the grey legs. Although the only plumage colouring known today is black and white, an old print by a Dutch artist illustrates the fowl with an all-white plumage and no black marking whatsoever. Its closest relation is the Vorwerk, which is very much a man-made breed and is very rare in Britain, although it is more widely known in Europe.

In spite of the great attractions of the Lakenvelder, with its contrasting black and white plumage giving it a most striking appearance, and the cockerel, in particular, being most handsome, it never really caught the imagination of the poultry fanciers of Britain. This has resulted in the breed being a very rare exhibit at most poultry shows within the British Isles.

The introduction of the Lakenvelder to British soil has been rather intermittent over many decades and, at the time of writing this book, there are only two poultry fanciers who are known to be breeding this fowl.

The cock has a most graceful appearance and a sprightly gait with a fairly long body and broad shoulders and a full and well rounded breast. The back slopes gently to the tail which has a good supply of well curved sickle feathers; although long, these hang quite gracefully, giving it a well balanced look. The wings are fairly large for the size of bird and are well tucked in to the sides.

The beak is rather short and quite stout for the size of bird. The single comb is straight and stands firmly erect; it ends abruptly with a flat rear edge and extends well back over the head. The ear lobes are almond-shaped and white. The wattles are well rounded and bright red in colour, as are the face and comb. The neck is not too long but is abundantly covered with softly flowing hackle feathers, extending well over the shoulders and breast. The legs seem to be a little short for the size of bird but the feet have reasonably long toes, well spread out, which give it a very firm stance. Both legs and feet are pale slate-blue.

The hen is similar to the cock, except that the tail is not quite so long and the comb is smaller.

Fully mature males reach weights of only about 2.2 kg (5 lb) and the female 1.6-2 kg (3½-4 lb). The eggs can be white or slightly tinted.

The colour of the hackle and tail feathers is a rich deep black, contrasting well with the remainder of the plumage, which is completely white.

Langshan — or Croad-Langshan (Plate 21)

The Langshan is an Asiatic breed of poultry which was first imported into England by the late Major Croad, hence its title. One of the qualities of this breed which made it so popular was the brown eggs it laid. The colouring of the egg shell was chocolate brown, which was much darker than any other egg laid by any of the other breeds of fowl in the British Isles at that time. This was the case until the Barnevelder breed, which also lays chocolate-coloured eggs, was imported into Britain in the early 1920s.

In days gone by, a large proportion of the population believed that brown-shelled eggs contained more nutrients, vitamins and proteins than white-shelled eggs and this assumption encouraged breeders to produce a breed of fowl for the colour of the egg rather than the quality of the bird. Even today, people still mistakenly believe that the brown-shelled egg is superior to the white.

There are, however, some strains of the Langshan breed that do not lay quite such chocolate-brown eggs as the earlier strains. This is due mainly to new blood being introduced through cross-breeding. Because of this cross-breeding, there are now two types of the Langshan breed: the Croad-Langshan and the Modern Langshan. The Croad-Langshan, when first introduced into Britain, was a rather short-bodied, thick-set bird. As a result of cross-breeding, the new strain of Modern Langshan was produced, and this has a slightly longer, tapering body, which is well cushioned and with a much straighter back.

The Langshan is classed as a large and heavy fowl but rather slow to mature. However, when fully grown, at say 6 months or so, it can reach weights of around 4.5-5.4 kg (10-12 lb). Its flesh is very similar to that of a turkey.

Like the Brahma and Cochin, the Langshan also has feathered legs but the feathering is not quite so prominent as it is on the outside of the leg only and is much shorter.

Although large, it is quite a graceful bird, well balanced with a rather proud and bold appearance, and it can be quite intelligent.

The colour of the surface feathering of the cock and the hen is black with a beetle-green sheen, free from any other colour, with the under plumage being a dark grey. Surprisingly, the toenails are white.

Leghorn (Plate 22)

The ancestors of the Leghorn came from the northern Mediterranean region, being first produced in the Leghorn district of Italy. It was, however, not until the early 1870s that the White variety was introduced into Britain from the USA. For almost 100 years, the Leghorn breed was considered by the majority of poultry breeders to be indispensable, because of its prolific and profitable egg-yield. Unfortunately, over the past years, it has fallen out of favour with the commercial breeders as a result of the development of commercial hybrids for egg- and meat-production. Because of its high egg-yield, the White Leghorn has, however, played a major role in cross-breeding and the development of some of the commercial strains.

Birds which are hatched in April are recognised as being good winter layers but, even so, discard the folklore idea that they do not moult — it is just not true. Every bird, regardless of breed, loses its feathers twice during the first year. The chicken first changes its feathers when it is approximately 4 months old and then again before a full 12 months are up. This second moult can, however, be delayed by adopting intensive feeding methods for a short period before the onset of the moult, but it certainly cannot be prevented.

It is far more advantageous to obtain the early-bred pullets because once they have started to lay their egg-production can be controlled and maintained much more easily during the marked change from summer to autumn and in the following inclement weather conditions. This is because they have had the opportunity to develop during the natural breeding season, i.e. spring. However, younger and undeveloped birds are not fully prepared for the climatic change and, as a result, can suffer a setback, which delays them coming into a satisfactory egg-laying condition, even if they are supplied with the high-protein balanced feed, which is so necessary for the development of the egg-producing organs. These must be well-developed to enable the hen to produce good quality eggs — always essential for hatching purposes.

During the 1920s and 1930s, the Leghorn was a very popular breed of poultry, especially with the commercial breeders, and more and more domestic poultry keepers changed to the Leghorn breed because of their high egg-yield and also because of the brown-tinted eggs which they laid. After their first egg-laying cycle, i.e. 14 months, they also proved to be acceptable table birds. In all, the Leghorn was found to be a very profitable breed of poultry.

Even today there is no doubt whatsoever that the general public still prefers to purchase brown-shelled eggs and, because of this, it can only be assumed that, although the flavour of the white-shelled eggs is equal to that of the brown, the coloured egg is more appealing.

White Leghorn

The birds should be pure white and free from any other colour or shading. The beak should be yellow and the single comb, eyes, face and wattles a bright red. The ear lobes should be white and the legs and feet yellow.

Fanciers who are interested in breeding White Leghorns for exhibition purposes should remember that there will always be a few of the pure-bred birds with a tendency to throw back to the colouring of their progenitors. This is a result of their being bred from the common Italian breed of Leghorn, which produced not only the pure White, but also the Black, Blue, Buff, Brown, Pile, Silver Duckwing and other variations of the breed. One should always bear in mind the possibility of a reversion to the colour of the progenitors in all pure-bred fowl, and poultry keepers who either hatch their own eggs or buy in a few day-old chicks might well be concerned to find a few of them developing the odd black, blue or buff feathers.

For exhibition purposes, it should be remembered that the comb of the cock should stand erect and not droop over one side of the face as happens with the hen bird. The legs and feet in both sexes should be yellow, sometimes with an orange tinge.

Black Leghorn

Black Leghorns, when produced from good-laying stock, are some of the best pure-bred egg-layers in the business. It must, however, be remembered that in every variety there are good and bad strains; it is essential, therefore, that every care should be taken in selecting the better layers if they are to be used as breeding stock. This applies to the production and development of all pure-bred fowl, especially when breeding for exhibition. Erratic and carefree breeding has, over the years, proved to be the cause of many disappointments. This problem can be greatly alleviated today, when purchasing eggs or day-old chicks, by buying in specimens that have been carefully selected on their history as good layers from reputable recognised breeders.

Several decades ago, the Black Leghorn was generally recognised as a utility fowl and, in the 1920s, a few Leghorns were produced which were completely black throughout their plumage and had yellow legs. Since then, of course, the fanciers and true breeders have taken pains to produce near perfect specimens of the Black Leghorn.

For exhibition purposes, the plumage of the cock and the hen should be rich black with a bluish tinge and absolutely free from any other colour.

Brown Leghorn
In the cock, the hackle should be of a rich orange, attractively striped with black, with bright red feathering at the front of and below the wattles. The shoulders and upper part of the wings (wing bows) should be of deep crimson with a slight reddish blending. Maroon is also acceptable. The bow coverts should be rich blue with a greenish sheen and form a bar across the wing. The secondary feathering should be brown (light or bay shading) and the primaries brown. The saddle should be a rich orange with a reddish tinge and with several black stripes. The breast and underbody should be shiny black, free from any other colouring. The tail should be greenish black, the coverts being black with a brown edging.

The hackle of the hen should be a rich golden yellow, attractively striped with black markings. The breast should be orange-red (salmon), deepening to maroon towards the head and light grey at the thighs. The body colouring should be of a rich warm brown, neatly compact, with black pencil markings. The tail should be black with brown pencilling.

Buff Leghorn
The plumage of both cock and hen should consist of any shade of buff but any sign of washiness or insipidness must be avoided.

The Malay, like the Asil, is one of the oldest breeds of poultry in existence, its history dating back far beyond the Roman era. Although the Malay was at one time indirectly related to the Asil, it does not have the same fighting instincts. The ancestors of the Malay that we know today were closely linked with the jungle fowls that lived in the forests and lowlands of the Asian continent. Like the Asil, they too were admired by the rulers of those Asian lands, not for gaming but for meat-production, to help feed the many influential visitors to those shores.

During the course of evolution, the breed continued to develop as a result of natural cross-breeding with other wild fowl. However, it was not until the early nineteenth century that, according to the records, they first set foot on English soil. Nevertheless, it is possible that the birds may have been imported from India by soldiers returning to England on leave. This would have been during the reign of Queen Victoria, before she was asked by the Indian rulers to become their Queen.

As it was a fowl of unusual shape, compared with the standard domestic hen of the early nineteenth century, poultry fanciers became more and more interested in the breed and it was at that time that crossing the Malay with the Old English Game became increasingly popular with the amateur geneticists, especially in Devon and Cornwall. As a result of this interest in cross-breeding, the Modern Game breed came into existence.

It soon became apparent, however, that the newly imported fowl was not going to be readily accepted as a general domestic breed, not just because of its pugnacious look but also because it laid very few eggs during the course of the year. Nevertheless, there are exceptions and some of the Malay hens bred today have much improved laying qualities. They do, however, produce a good quantity of breast meat and Malays have been known to reach weights of up to 5.4 kg (12 lb) or so.

Since the rejuvenation of interest in the breed, new strains have been developed, such as the Black, Black-Red, Pile, Spangled and White. Of all the different strains, the Black-Red is still the most common.

The cock stands very erect, almost upright on occasions. The head is broad, strong and powerful, with a fearsome expression and staring eyes. The comb is of the walnut type and set on the front of the head.

119

The rather long neck has very short feathering, almost non-existent towards the base. The body is quite wide at the front, tapering towards the tail, and has a full and deep breast with very little plumage. Wings are large and powerful, positioned high up on the body and tucked well in. The tail is very narrow and slightly drooping. The legs have long muscular thighs, again with very little plumage. Shanks are also long and evenly scaled and the four toes are long and powerful.

The tail of the hen is carried much straighter than that of the cock and the attitude of the body is not quite so upright; otherwise, the general characteristics are similar to those of the cock, apart from the natural sex differences.

Maran (Plate 24)

This breed takes its name from the town of Marans in France. It is quite a cosmopolitan bird, having in its make-up such breeds as the Croad-Langshan, Faverolle, Plymouth Rock (Barred) and several others. Although the Maran has been in existence for some time, it is a comparative newcomer to the British Isles as it was not introduced to the fanciers until as late as 1926-28. One of the reasons for its popularity was the barred effect on the plumage, which is most striking and quite attractive. By the early 1930s, the breed was even more popular, because it was becoming obvious that it had good sitting qualities and also good parental instincts. As time passed by and World War 2 began in earnest, other strains were developed, e.g. the Black, Dark Cuckoo and Silver Cuckoo.

The Maran is by no means a lazy bird. It has a quick movement and is always very active, never last on the scene when tasty morsels of food are being handed out. Sadly, because of this agility, it does not readily make a very good show bird. Nevertheless, it is most certainly an acceptable utility fowl; it is classed as a heavy breed and lays darkish brown eggs. A fully mature cock can weigh up to 3.6-4.5 kg (8-10 lb) and a hen about 2.7-3 kg (6-7 lb). It is a fine table bird producing succulent white meat.

In all the various colour combinations, the hens are similar to their cocks, in both colouring and form.

Dark Cuckoo
As the name implies, the plumage is a cuckoo pattern throughout, with each feather being barred with a black or bluish black band,

regularly spaced along each feather, giving an appearance of gentle ripples. The plumage is always neat and compact and the wings well tucked in.

Golden Cuckoo
This is very similar to the Dark Cuckoo in plumage pattern but with obvious differences, such as the hackle feathers having a blue-grey colouring with golden and black bars. The breast is also shaded a blue-grey with black bars blending to a golden shade towards the upper section of the breast. Wing bows and shoulders are bluish grey, shaded with black bars and sometimes with golden edging.

Silver Cuckoo
In the Silver Cuckoo, the neck plumage is mainly white and the upper section of the breast has similar colouring. The remainder of the plumage is barred throughout, perhaps a shade lighter than the Dark Cuckoo.

Minorca (Plate 25)

The Minorca is of Mediterranean origin and was first imported into the British Isles from the island of Minorca in 1830. It soon became quite popular because of its stocky build and the large eggs which it laid. It can be classed as the largest of the Mediterranean breeds, somewhat heavier than the Ancona, Andalusian and Leghorn. Today the breeding and development of the Minorca is mainly in the hands of the ardent fancier and enthusiast who breeds principally for show. The breed has altered in size greatly since its introduction and is nowadays quite different to the small compact birds that were originally imported into Britain.

In the earlier years of development, the Minorca and the Old Spanish Fowl looked very much alike, because they were of almost the same build, but as their development proceeded the Spanish Fowl became almost extinct, which is not surprising as the size of the head became completely abnormal and looked quite grotesque; in some cases the chicks were born blind.

The development of the Minorca has continued to progress steadily over the years, because of the dedicated attention to cross-breeding by the ardent fanciers. As a result, the type of Minorca breed that we see today is a bird which is pleasing to the eye, with its shiny black plumage, white ear lobes and brilliant red comb and wattles — colours which contrast greatly with each other.

Because of their compact neatness, the birds, when fully mature, look very attractive. The drooping of the brilliant red comb, which has a tendency to hang forward and to one side of the face, gives them a flamboyant air which is quite eye-catching.

Although the Minorca is classed as a large fowl, it is by no means one of the heavy breeds of poultry. It does not put on flesh at the same rate as the heavy breeds because of its incessant activity, and it is not therefore a fowl to keep in close confinement.

Over the years, apart from the Black Minorca, the other strains which have emerged are the White and the Blue. For showing purposes, any variations in the specifications given below can only be classed as serious defects.

Black Minorca

The plumage of the cock and hen is glossy black and, in both, the main features of the head are the dark eyes, blood-red comb, face and wattles and pure white ear lobes. The legs and feet are black and the beak and toenails are a dark horn colour.

White Minorca

The plumage of both sexes is a glossy white. Eyes, comb, face and wattles are a brilliant red. The ear lobes are pure white, and the legs are also white with a hint of pink.

Blue Minorca

Both cock and hen have plumage which is light to medium blue and free from any other colouring. In the cock, the hackles, wing bows and back can be a slightly darker shade. Legs and feet are blue (slate). The features of the head are the same as for the Black.

New Hampshire Red (Plate 26)

The farmers and poultry fanciers of the American state of New Hampshire have a lot in common with their counterparts of Rhode Island, for it was the Rhode Island Red fowl that figured strongly in the creation of the New Hampshire Red. This breed was produced as a result of the selective breeding and in-breeding of the Rhode Island Red by the dedicated fanciers and poultry breeders of New Hampshire.

One of the reasons for the development of the new fowl was to try to produce a bird which would have improved qualities, i.e. with a larger frame and better meat-production, as well as being a prolific egg-layer, together with a more attractive plumage. However, although the breed is quite different in many ways to that of the Rhode Island Red, including the shape of its body and the lighter colour of its plumage, it never really attained great popularity. Nevertheless, this most attractive fowl, with its fine qualities and shining feathers, is very pleasing to the eye. It can most certainly be classed as an ideal table bird , as the cock can reach weights of up to 4 kg (9 lb) when fully grown and the hen about 2.7-3 kg (7 lb). However, its egg-production does leave something to be desired when compared with the Rhode Island Red.

It was first introduced into Britain in around the early twentieth century, when the poultry fanciers decided to take a hand in its further development. As the years have rolled by, without crossing with any other breed, the strain which exists today now has laying qualities that are much improved, as well as being quick to reach maturity. Even so, it has never really created much enthusiasm with the general poultry farmers of today. The Rhode Island Red cock is still the more widely used fowl for breeding and cross-breeding by the commercial hatcheries, in order to produce the hybrid strains that supply the ever-popular brown eggs.

This is a large and heavy breed. The cock is active and well proportioned with a full and well rounded breast, somewhat deeper than that of the Rhode Island Red. Its wings are of average length and well tucked in. The tail and sickle feathers are also of medium length. The eyes are fairly large and quite prominent and the single comb is of medium length with six points, five of which are well defined. The wattles are of average length, reasonably smooth and flat. Ear lobes are elongated, almost touching the wattles. The neck is of medium length, with a more upright attitude than that of the Rhode Island Red, and generously covered with hackle feathers. The legs and feet are well proportioned. The thighs are average but certainly larger than those of the Rhode Island Red, and the four toes are quite straight and also well spread apart.

The hen is similar to the male, apart from the normal sex differences. Eggs are a light or tinted brown colour. Weights are similar to those of the Rhode Island Red: 3.6-4 kg (8-9 lb) for a mature cock and 2.7-3 kg (6-7 lb) for a well developed hen.

The whole of the cock's plumage consists of various shades of rich light and deep chestnut-red shades. The head is covered with the lighter shade, blending into the slightly deeper chestnut-red of the neck and breast. The richly coloured saddle feathers are again slightly darker than the neck shade. Wings are a mixture of the chestnut-red shades: the wing fronts are medium in colour and the bows are more brilliant. Coverts, primaries and the upper web are medium in colour but the lower web and primaries have a black edging. Tail feathering and sickles are a very dark shade, almost black with a hint of beetle-green lustre.

The plumage of the hen is mainly of a medium chestnut-red throughout, the main difference being that each feather covering the head and neck region is tipped with a lighter shade. Other areas of the plumage, such as the lower neck, coverts and main tail feathers, are all edged with a black trim.

Old English Pheasant Fowl

The Old English Pheasant Fowl is most probably the oldest of all the pure English breeds in existence along with the Old English Game. It was certainly a very popular bird in the northern counties of England for it was bred by many of the farmers, some for domestic reasons, but the majority for gaming. The coal-miners of Durham and Northumberland also shared an interest in the breed, but this was purely to provide themselves with an abundant supply of fresh eggs and poultry meat to subsidise the family budget.

It was not known as the Old English Pheasant Fowl until the beginning of World War 1, before which it has been known for many years as the Old-Fashioned Pheasant or Pheasant Fowl. In Yorkshire, it was also known as the Yorkshire Pheasant.

In 55 B.C., when the Romans invaded Europe, news had already reached Julius Caesar of 'the barbaric English pastime of cock-fighting' and it was shortly after the Roman invasion of Britain that further details of these 'pastimes and pleasures' were recorded in his memoirs. There is a long history of 'cocking' in Britain and, even at the present time, game-shooting is still practised by the nobility and gentry of the land, as well as by farmers and country folk.

The breeds of Old English Pheasant in existence today consist of the Gold and Silver and, although it is now a rare event for the eye to behold, a flock of these birds in a country setting is a most awe-inspiring sight.

Being of game descent, the breed is always alert and very active. It

124

has a rather long body with broad shoulders and a well-rounded breast. The tail stands well out at an angle of 45° to the line of the back. One of the bird's main features, apart from the colourful plumage, is its rose comb, which ends in a long tapering point. The cock and the hen are very similar in make-up, apart from the natural sex differences and the tail. The tail of the cock has well-shaped sickle feathers which stand well up and hang quite gracefully.

The Old English Pheasant Fowl is not a heavy breed and can reach weights of only 2.2-2.7 kg (5-6 lb) for the male and 1.8-2.1 kg (4-5 lb) for the female. The eggs are white.

Golden
Both cock and hen of the Golden variety have a ground colouring of rich bay or light buff and the back of the cock is a rich mahogany red. The tail is black with a prominent beetle-green sheen. The hackle feathers have a striped effect and are lightly tipped. Saddle feathers are of a slightly darker mahogany shade than the neck area.

Silver
Of the Silver variety, the cock and hen have almost the same plumage colour, i.e. white with black markings that have a slight beetle-green sheen. The legs and feet in both sexes are a slate-blue colour, which indicates dark flesh.

Orpington (Plate 27)

The Orpington breed is a very hardy strain, which is suited to almost every condition under which poultry can be kept, with the exception of the battery cage, because of its size. This pure English breed is classed as a large fowl and was named after the Kentish village of Orpington, where the well-known poultry breeder, William Cook, had his farm and hatcheries. The first strain of this new breed was the Black Orpington, which was first bred in the year 1886. Then two years later, William Cook, with his dedicated interest in the domestic fowl, developed the White Orpington. He followed this, again two years later, with the Buff Orpington. During the next seven years, there was very little progress made in the further development of the strain, until the year 1897, when the Jubilee Orpington emerged. Then followed the Spangled in 1900.

Of all the strains produced, in the author's opinion the Buff is still

125

the most attractive of the Orpington breed, not only because of its rich colouring but also because of its character: for, when confronted with other breeds of similar size, it behaves in a docile manner, cowering down with its head almost touching the ground.

The Buff is the third generation of the Orpington breed and is an excellent table fowl as well as a good layer. In fact, it will produce almost as many eggs in the second year as it does in the first, providing that it is fed correctly with the standard balanced layers' mash or pellets. With all the heavier breeds, it is essential to avoid over-fatness after reaching the end of the first laying cycle, i.e. at 14 months old. The Buff Orpington is undoubtedly one of the most attractive as well as one of the most profitable breeds in terms of egg- and meat-production, if it is purchased from a reliable source.

The breed soon became very popular, not only because of its laying qualities but also because of its enormous size. A fully mature male bird can reach a weight of some 4.5-6.4 kg (10-14 lb) and the hen 3.6-4.5 kg (8-10 lb), which makes it an ideal table bird. Apart from its weight, it also lays brown eggs, which are still, as has been previously mentioned, considered to be more nutritious, even today.

By cross-breeding, fresh blood can be introduced into a strain, resulting in stronger, healthier offspring with plumage of greater colour variation. This was the case with the original Orpingtons, which were eventually crossed with the Cochin and Croad-Langshan, the Plymouth and Minorca breeds. The resultant offspring were of enormous size with a much more compact plumage. However, because of this development, the egg-production suffered to such an extent that, today, the Orpington is more of a show bird than a breed for utility purposes and general domestic poultry keeping.

The skull is quite small in relation to the body, with a strong, well curved beak. The eyes are full and bright. The single comb is quite small, but straight and standing erect. The face is smooth and the ear lobes are elongated. Wattles are rather small and oblong and the rather short neck is well covered with plumage. The body is deep and quite broad with a well rounded breast. Surprisingly, the wings are quite small in relation to the size of the bird and the ends are hidden under the saddle feathering. The tail feathers are rather short but stand up quite well. The legs are short and stumpy, strong and well spread apart. The thighs are almost hidden by the body feathers and the four toes are straight and well spread out.

The hen is generally similar to the cock but the tail has a more flattened look and lies at an angle of about 45° to the body, giving it a more graceful appearance.

126

Both cock and hen Orpingtons have an abundance of plumage which is reasonably soft and fluffy. The Orpington is classed as a heavy breed. The eggs are brown.

Plymouth Rock (Plate 28)

The Plymouth Rock or Rocks is one of the largest of the poultry families and dates back to the early nineteenth century. The Rock is an American breed, which originated in the town of Plymouth, New Hampshire, home of the New Hampshire Red. Members of the family are the Black, Blue, Barred, Buff, Columbian, Canadian Barred and the White. It was not until about 1870 that the Barred Rock was first imported into Britain, after which, following closely on its heels, came the Buff Rock. There were also the Silver-Pencilled and the Spangled Rocks, but very little is heard of these two strains nowadays. The breed is classed as a heavy fowl.

The Rocks lay brown-tinted eggs and are considered to be above-average layers. Being a large fowl, they are also readily acceptable as good table birds, with tender succulent flesh. A cockerel can weigh between 3 and 4 kg (7 and 9 lb) and is sometimes heavier. A hen can produce a weight of some 2.7-3.6 kg (6-8 lb) when fully mature.

The Plymouth Rock is widely used in cross-breeding with the Rhode Island Red and Sussex to produce commercial hybrid strains, either for meat-production or egg-laying. Such hybrids are vital to the nation's food supply.

Barred Plymouth Rock
This, undoubtedly, is one of the most handsome of the breed, the base colouring being white with a bluish tinge, barred with a black/beetle-green sheen, which has the effect of giving the plumage a bluish appearance. Each feather should finish with a black tip, as well as having a barred effect.

Columbian Plymouth Rock
Both the cock and hen of the Columbian Rock are very similar in appearance to the Light Sussex, which could be the result of using similar progenitors in the development of the strain. The plumage is pure white with the hackle striped with a black/beetle-green and laced with white. The tail is of a similar shade. The legs and feet are yellow in colour.

Buff Plymouth Rock

The Buff Rock plumage gives a very pleasing effect to the eye and, like that of the Buff Orpington, is quite attractive in appearance, being a rich golden buff in colour. The head and feet are yellow, giving quite a striking effect and blending extremely well with the rich plumage. Because of its make-up, the Buff Plymouth Rock has a tendency, periodically, to produce offspring that develop feathering on the legs; this leg feathering is the result of two of its progenitors being the Cochin and the Brahma.

White Plymouth Rock

The White Plymouth Rock has a plumage of pure white in both cock and hen. The breed was first produced in the late nineteenth century in the State of New Hampshire, also the home of the Barred Plymouth Rock. Because of the pure white colouring, it is rather difficult to keep the birds in an attractively clean state, unless they are kept in a controlled environment. Their yellow, or sometimes orange, legs and yellow beak make them truly attractive birds.

General characteristics of both the cock and the hen are very similar to those of the Sussex, with the exception of the variation in colouring.

Poland (Plate 29)

Ulysses Aldrovandi, who was born in 1527 and lived for some 78 years, was a great poultry enthusiast and it was he who named this breed the Poulander or Paduan Fowl over 300 years ago. This is substantiated by the discovery of historical writings and old drawings of the fowl that were found in the region known as Padua many decades ago. However, a great deal of confusion still surrounds this ancient and fascinating breed, as it also became known as the Poland, Polish and Pole Fowl.

Although it is fairly certain that the fowl did originate in the Paduan or Patavinian region, it is most probable that another strain also existed in other European countries at the same time. Holland is one country that is predominant in its history, for it was the Dutch poultry fanciers who bred their own strain of Polish Fowl, which was known as the Polish or Poland. The name Polish, when referring to the breed of poultry, is derived from the word *polle* which is Dutch and Low German for 'head', and in this context means 'hornless' or

Plate 1 (right) Ancona cock. Always alert, and with a bold appearance, this bird is usually very active.

Plate 2 (below) Andalusian cock and hen. This is one of the most attractive of all the Mediterranean breeds. Notice the distinct black lacing on each feather.

Plate 3 (bottom) Silver Appenzeller cock and hen. Each feather ends in a distinct black spangle and the head is adorned with an unusual crest or 'bonnet'.

Plate 4 (above) Lavender Araucana cock. This variety has a delicate lavender-coloured plumage and a rose comb. The hen lays pale blue or green-tinted eggs.

Plate 5 (right) Red Asil cock and hen. This cock is making amorous advances to his mate. The Asil is one of the oldest known breeds of Game Fowl and will always fight to the death.

Plate 6 Partridge Barnevelder cock. The Black variety of the Barnevelder is almost extinct. The breed is often referred to affectionately as 'Barney'.

Plate 7 (above left) Golden Brahma cock.

Plate 9 (below) Buff Cochin.

Plate 8 (above right) Silver Campine cock.
Plate 10 (bottom) Silver Grey Dorking cock,
guarding his hens.

Plate 11 (top) Salmon Faverolle hen.
Plate 12 (below left) White Frizzle cock.
Plate 13 (below right) Jubilee Indian Game cock with a Dark Indian Game hen.

Plate 14 (below left) Large Modern Game cock of the Silver Duckwing variety.
Plate 15 (below right) Old English Game cock (Black-Breasted Red, Oxford type).
Plate 16 (above) Gold-Pencilled Hamburgh cock.

Plate 17 (top) Houdan cock.

Plate 18 (above) Red Jungle Fowl cock.

Plate 20 (above) Lakenfelder cock. This is a beautiful specimen of the breed.

Plate 19 (top) La Flèche. This is a very old French breed.

Plate 21 (right) White Croad Langshan. First imported into Britain in about 1872.

Plate 22 (above) Black Leghorn hens on free-range. This breed was most popular as a domestic fowl in the 1930s because of its prolific egg-laying qualities.

Plate 23 (left) Malay cock. The bedraggled appearance of the Malay is quite normal.

Plate 24 (below) Dark Cuckoo Maran cock and hen. The hens make excellent mothers and are most suited to natural breeding.

Plate 25 Minorca hen. This is a perfect specimen of true Mediterranean stock.

Plate 26 New Hampshire Red cock.

Plate 27 Buff Orpington cock. A pure English breed, this was named after the Kentish village where it was first produced. It has a surprisingly small head for the size of body. The cock can reach weights of 4.5 kg (10 lb) or more.

Plate 28 (top) Barred Plymouth Rock cock and hens. This was first bred in the United States of America in about 1850. Each feather has black bars along its length and ends with a black tip.

Plate 29 (above) Chamois Poland cock and hens. These beautiful birds with their highly decorative plumage are undoubtedly one of nature's treasures.

Plate 30 (left) Derbyshire Redcap cock. A favourite with the miners of Derbyshire, this was first bred in the early 1900s. Its main attraction is the enormous rose comb.

Plate 31 (top) Rhode Island Red hen. This breed, together with the Light Sussex, was among the most popular of general-purpose breeds in Britain during the 1920s and early 1940s.

Plate 32 (above) Sicilian Buttercup cock. This is another of the Mediterranean breeds. This portrait shows the saucer-shaped comb which is said to hold rain-water so that, in very hot climates, after rainfall, the birds can drink from each other's combs.

Plate 33 (right) Silkie or Silk Fowl cock. The plumage is very unusual, almost hair-like. A small fowl, it is by no means a bantam.

Plate 34 (top) Spanish cock and hens. At one time this was one of the best layers of large white eggs.

Plate 35 (above) Sumatra cock. An Asiatic breed, which reached the shores of Britain in about 1900, this dark, strange and mysterious creature is very rare. The hen is a prolific egg-layer and a good sitter.

Plate 36 (right) Speckled Sussex cock. The Speckled is said to be the oldest strain of the Sussex, which used to be known as the 'Kent Fowl'.

Plate 37 (above) Warren Studler hens with a Maran hen. The Warren Studler is a commercial hybrid strain produced solely for laying brown eggs.

Plate 38 (below) Welsummer cock and hens. This Dutch breed was named after the village of Welsum in northern Holland.

Plate 39 (above) Silver-Laced Wyandotte cock and hens.

Plate 40 (below) Red-Saddled Yokohama cock.

'without a comb' (as in poll-cattle). Therefore, we now know that at least two different strains of Poland were in existence during the same period: one having a horn-type comb, known as the Poulander; the other being 'hornless' or virtually without a comb, and called the Pole Fowl or Poland.

Regardless of the above, confusion still reigns with some poultry fanciers and breeders, as they are also of the opinion that the breed may possibly have originated in one of the eastern countries. However, there is no evidence to substantiate this belief. Although many origins have been suggested for the Poland, there is little evidence to indicate that this crested breed actually hails from the country of Poland, as its name might imply.

The Padua or Poland comes in Self colours, i.e. Self White, Self Black and also Self Blue. The Blue has not yet been standardised in Great Britain. There is also the Chamois, Gold-Laced and Silver-Laced. These three strains all have blue legs and feet, as well as beards.

The White-Crested variety is a direct descendant of the Dutch breeds of Poland or Polish, and these are known in Britain as Dutch White-Caps or White-Crested Dutch. There is the White-Crested Black, the White-Crested Blue and also the White-Crested Cuckoo. These White-Crested varieties do not have beards, but they do have wattles. Neither do they have blue legs and feet, but shaded horn-coloured legs and white feet, this feature being the difference between the two types. Although they are in fact two different varieties of fowl, in the British Isles they are all referred to under the one name 'Poland'. In Europe, the muffed and bearded varieties are still referred to as Paduans.

The Poland fowl is a very rare species and by far the greater number of birds in existence are in bantam form.

In the cock, the curiously dome-shaped skull tends to emphasise rather than increase the crest size. The back is quite long and rather flat. The body has a full but moderately rounded breast, with wide shoulders and deepish flanks. Wings are of reasonable length and well tucked in. The tail is neatly shaped and carried at an angle of about 45°. The body is covered with an abundance of plumage. The predominant feature of the breed is the large white crest of plumage on top of the head.

With the exception of the crest, which is round in shape, the hen is similar to the cock. They are considered to be good layers when kept in their right environment. The eggs are white.

Redcap (Plate 30)

The Redcap, sometimes called the Derbyshire Redcap or Yorkshire Redcap, is a delightful fowl by any standards, so much so that its showing qualities extend back as far as the early nineteenth century. It is one of the oldest of the very few pure breeds of England, which at some stage in its history must have been crossed with the true Old English Game Fowl, as the colouring of the plumage is very similar.

In the late eighteenth and early nineteenth centuries, it was an extremely popular breed with the Derbyshire quarry-workers and lead-miners, who knew it to be a prolific egg-layer as well as a good meat producer. It was also recognised as a hardy fowl that could stand up to the very cold winters experienced in the Derbyshire Peak district and Yorkshire moorland. Being a very good forager, it was a breed which required little attention, and was therefore partly self-supporting. Because of these excellent qualities, poultry fanciers in many other countries, such as the USA, Australia, Canada, Germany, Holland and New Zealand, have shown a great deal of interest in the breed.

The Redcap is still one of the most typical of the northern breeds of England and, during the course of its evolution, it was also known as the Manchester Pheasant Fowl or Moss Pheasant. Obviously, there were slight differences in their make-up, but the appearance of these birds points to a very close relationship with the Yorkshire Redcap. The main features are its bright red ear lobes and very large rose comb. At one stage in its development, emphasis was placed on increasing the size of the comb, which eventually became very large and grotesque. Unfortunately, this development led to a general loss of the breed's utility merit.

The most prominent feature of the cock is its very large rose comb, which in a good specimen, is evenly bobbled and ends in a straight spike or leader. The comb covers the entire head but is well raised off the eyes and beak. The ear lobes and wattles are well proportioned in relation to the head and are bright red in colour. The neck is slightly arched near to the head and covered with an abundance of hackle. The back is rather broad, moderately long and ending with beautifully arched sickle and side-hangers, with a flowing appearance, like the breaking of a large wave. Wings are of average length and well tucked in to the body. Legs and toes are set wide apart, giving it a very firm stance that supports the bird well in high winds.

The general characteristics of the hen are the same as for the cock, the only exception being, apart from the normal differences, that the rose comb is almost half the size of that of the cock.

The Redcap is by no means a heavy breed as it can reach weights of only 2.2-2.7 kg (5-6 lb) for a fully developed male and 1.8-2.3 kg (4-5 lb) for a good female. The eggs are white.

The neck hackle and saddle feathers of the cock have a beetle-green webbing effect, with very fine black tips. The back, wing bows and coverts are a rich red and there is a black barring across each wing. The primaries and secondaries are red with black. Breast and keel are also black.

In the hen, the colour of the hackle is the same as in the cock but the back and breast is a deep, rich brown, each feather having a black spangled effect. Other points are the same as for the cock.

Rhode Island Red (Plate 31)

Since its introduction into the British Isles in the early twentieth century, no breed has made such rapid progress in its development and popularity as the Rhode Island Red. The breed was developed in the American state of Rhode Island and was the result of crossing the Malay, Java and Indonesian breeds with the Black-Red Oriental strains from China. Of course, it took many years to establish a firm colouring for the breed. However, by the 1930s, a true depth of colouring and richness became apparent in the new strains and, in the late 1930s and early 1940s, the Rhode Island Red was considered to be a highly successful breed, both for its laying qualities and meat-production.

There were a good many poultry farmers in business during the pre- and and post-war years who would readily agree that this breed was one of the most profitable general-purpose fowl available at the time, because of its size and fitness and laying properties.

When buying in stock of Rhode Island Reds, or of any breed, it is advisable to visit the breeding farms, to ensure that the condition of the stock is satisfactory. Many poor strains can still be found as a result of inexperienced breeders not taking the trouble to eliminate unsatisfactory hens but continuing to use them for breeding purposes. Disappointment can be avoided by taking care to ascertain the pedigree of the birds and, if possible, making your own selection from the stock. For those who require perfect strains, either for the show-bench or breeding purposes, the value of each of the progeny should be taken into consideration. When buying in any true breed of poultry, it is most important to make sure that the birds taken into stock are fully representative of the breed. This applies to all two-coloured breeds and not just to the Rhode Island Reds.

Even today, the Rhode Island Red is still an important breed, as it is widely used in breeding by many commercial hatcheries in order to produce the brown-feathered hybrid strains that supply the ever-popular brown eggs. It is also crossed with the Plymouth Rock and Sussex breeds to produce poultry with a genetic structure that enables the fowl to develop breast meat in a much shorter period than normal.

The head should be of medium size, reasonably stout but not too thick. The beak, which is slightly curved, is also of medium size. The eyes are rather large, bright and prominent. The comb is either a single or rose, the single comb being quite firm and upright with five well-defined serrations. The surface of the rose comb is covered in small points, finishing with a small spike at the rear, which curves over the back of the head. The face is smooth and the ear lobes well developed and of a fine texture. Wattles are of average size, reasonably well rounded and with smooth skin.

The neck should be of medium length, protruding slightly forward and well covered with hackle feathers. The broad and long body has a deep keel bone that should be straight and the breast is broad and full. The body is oblong rather than square. The back is long, broad and almost horizontal but with a slightly rising curve towards the neck and hackle feathers. The wings are rather large and well tucked into the sides. The tail is somewhat small in relation to the size of the bird, but the feathering is well spread apart, carried slightly low, but by no means drooping and with a tendency to make the bird look much larger than it actually is.

The legs are well spread, giving the appearance of a firm stance. The rather large thighs are well rounded, of medium length and free of feathers. The four toes are of medium length and well spread apart. The cock weighs between 3.6 and 4 kg (8 and 9 lb). The eggs are light or medium brown.

The beak may have a reddish tinge but can sometimes be yellow. The eyes, comb, ear lobes and wattles are a rich red colour. The legs are yellow but can also have a reddish shading. In the cock, the neck hackle is a rich red, blending well with the breast and the back. Wing plumage is a blend of colouring with the lower section of the primaries (the web) black and the upper parts red. The lower section of the secondaries is red and the upper section is black. Flight coverts are also black but the bows are red. The main feathers of the tail should be black or beetle-greenish black. The remainder of the plumage and the general surface of the cock should be a rich red, free from any other shading, except for the black marking. In general, the variations in the colouring of the plumage should smoothly blend one into the other. The underfeathering should have a pale red or salmon shading.

Like the cock, the neck hackle of the hen is red with the lower feathers having a black tip but not with a prominent lacking effect. The tail is black or beetle-greenish black, but does not lay at the same angle to the back as that of the male; it is also a little shorter. In other respects, the cock and hen are similar.

Perfect specimens of both should have a glossy or brilliant sheen on their plumage.

Sicilian Buttercup (Plate 32)

As the name implies, it is accepted in some quarters that the breed originated in Sicily, where it was kept by the farmers and peasants as a general domestic fowl. However, its exact origin is still a mystery, because it is also thought to have been introduced from Egypt, parts of North Africa or other areas of hot climates in Africa. This assumption is based on the cup- or saucer-shaped comb. It is believed that, when it rained, because of the shape of the comb, the rainwater used to gather in the comb of each fowl. Thus each of the flock, in turn, could be supplied with fresh water, as they used to drink out of each other's combs. It is a tale which has not yet been proved; nevertheless it is quite possible because the comb will hold liquid.

The Sicilian Buttercup is a very flighty bird, because of its small frame and light weight. It is one of the wildest of the Mediterranean breeds and is very difficult to catch.

It was first imported into the British Isles in the early twentieth century, and one of the main reasons for its eventual popularity was its prolific egg-laying qualities and low food consumption. The breed is also very popular in the USA, where it was used to produce birds that were more attractive and with greater utilitarian qualities. In recent years, geneticists and fanciers have continued to experiment but have found it very difficult to produce a comb of similar design. One of the methods adopted is to cross a breed with a double comb, such as La Flèche, with one bearing a single comb. Eventually, offspring will be produced with a double comb, joined at both ends, forming a cup or saucer shape.

Because of the shape of its comb, the cock does seem to have a regal appearance. It also has a bold and upright stance and is very active. The body is fairly long and deep with a well rounded breast and a broad back, reasonably well curved. Its wings are rather long for the size of bird but well tucked in. This may be one of the reasons for its flightiness. The long flowing tail with well curved sickles tends to give it a graceful appearance.

The main feature of the head is the crown or saucer-shaped comb, sitting on a fairly long skull, by normal standards. The eyes are quite large and always with an alert attitude. The beak is of medium length and the wattles are thin but well rounded. The medium length neck has an abundance of hackle feathers draped well over the shoulders. The legs are of average length but are set well apart. The shanks are clean with neat scaling and the four toes are completely straight and well spread.

The comb in the hen is somewhat smaller and not quite so prominent; otherwise, the general characteristics are similar to those of the cock.

The different strains which have been developed over the decades are the Brown, Golden, Golden Duckwing, Silver and the White. The eggs are white and birds can reach weights of approximately 2.7 kg (6 lb) for the cock and 2.25 kg (5 lb) for the hen, or sometimes a little heavier, depending on the environment and standard of feed.

Silkie (Plate 33)

The Silkie is a most unusual fowl of oriental descent. The actual year of origin can only be a calculated guess as very little is known of the history of this rather ornamental breed. Many advocates believe that the fowl originated in China almost 1000 years ago; others are more convinced that a much stronger Japanese influence prevails in its genetics. In addition to this, speculation has also included India as its birthplace. Whatever the outcome of such speculation, there is little doubt that the Silkie is of Asiatic origin.

During the sixteenth and seventeenth centuries, the breed eventually found its way into Europe and at that time was known as the Silk Fowl, as the plumage was much softer and more hair-like than it is today. Because of this very soft, hair-like plumage, much care had to be taken to protect the young chicks for, if the feathers became damp or wet, there was always a danger of the youngsters being hanged or choked to death, as they became entangled in their bedraggled plumage.

The Silk Fowl or Silkie, as it is called today, was first introduced into the British Isles around the mid-nineteenth century and it was only then that the poultry geneticists and dedicated fanciers started to develop a strain that would have much stronger and broader feathers, but would still retain the silky and fluffy appearance. So, today, we have a fowl which is unique in its appearance and worthy of exhibition.

The Silkie is by no means a perfect utility fowl, as its egg-laying record leaves something to be desired. Being a small fowl, but by no means a bantam, its meat-producing quality is also poor, as it can reach weights of only 0.9-1.4 kg (2-3 lb), when fully grown. Although it has very little to offer as a general domestic fowl, apart from its ornamental qualities, the hens do make very good mothers, as they frequently develop a broody instinct, and this has always been their main function.

The Silkie cock has a rather broad body but a short back. The breast is also broad, with well rounded shoulders and a fairly short neck, abundantly covered with hackle. The tail is not too short, with feathers that are somewhat harder and tougher than the rest of the plumage on the body. The tail feathering, however, is much shorter and without the flowing effect that is significant with other breeds of poultry.

The head is small with a flowing crest that has a swept-back effect, the feathers being smooth and silky. The beak is stout but short. The eyes, which are full and alert, have an oriental or Siamese appearance. The comb is more unusual, almost round in shape with a groove or furrow across it from left to right. Ear lobes are oval and the wattles are almost circular or semi-circular.

The legs and feet are set quite wide apart in relation to the size of bird and are rather short. Like the Langshans and Brahmas, there is leg-feathering but the plumage in this case is very soft and silky. Even some of the toes have feathering, which indicates that the Silkie is not a fowl to be given freedom in damp and wet conditions.

The hen differs from the cock in some ways. Apart from the normal sex differences, the crest is shorter and more compact, the wattles are smaller and not quite so rounded and the comb is much smaller and flatter and completely different in design.

Since the introduction of the Silkie to the shores of Britain, other strains have been developed and these include the Black, Blue and Gold.

The original and true Spanish fowl is, of course, one of the oldest of the Mediterranean breeds, with a plumage of predominantly black and grey feathers, but with white ear lobes. In Spain, this type of fowl can be classed as the general domestic or farmyard hen, as it is widely bred by farmers and domestic poultry keepers as a utility bird, because of its large frame and because it is considered to be one of the best white-egg-producers in all the Mediterranean breeds. This particular type is truly a native of Spain, together with the Andalusian, the only difference being that the Andalusian was produced and developed in Britain, although all the genetic material was of Spanish origin. All the original Spanish fowl were of a uniform type and it was not until the introduction of the Red-Faced Minorca that the popularity of the Spanish domestic breed started to decline.

Of all the Black Spanish fowl, it is the White-Faced variety, with its spectacular appearance, that has been developed outside of its native country. Old writings indicate that the 'enormous' White-Faced Spanish was in existence even as early as the 1640s, shortly after the Shakespearean era. Most of their development took place in Britain and Holland, although other countries were also involved in experiments. However, it is in Holland that the development of the white face seems to have occurred. At one time, Holland, or an area of Holland, was a Spanish province and it was at about that time that people started to show an interest in poultry. Although nearly all the Spanish breeds have white ear lobes, it is only this strain that has a completely white face as well. The face and comb are separated only by a hair-line of black along each side of the comb and the rest of the plumage is completely black. Even today, this White-Faced variety is still not generally bred in Spain.

During the past century, Spain seems to have been a reservoir of heavier different types of deeper-bodied Mediterranean fowl. A great number of these Spanish domestic fowl were imported into the USA, where they were crossed with the local village fowl. However, it was in England, during the latter part of the nineteenth century, that many more experiments in the development of the White-Faced Spanish took place and, nowadays, there are more White-Faced Spanish in the British Isles than there are in Spain. Nevertheless, it still remains a very rare breed.

Although it is classed as a light breed, the cock is a fairly large bird, quite unusual in appearance because of its completely white face, which has the effect of giving it a sorrowful or weird look, as if made up like a clown.

The head is quite large with a bright red single comb that stands firmly upright. Ear lobes are long and broad and lower than the bright red wattles. The neck is moderately long and covered with an abundance of flowing hackle. Its body is fairly long, broad at the front with a less than full breast that has a rather flat appearance at the base. The back is sloping, with flowing coverts and sickle feathers, the latter hanging gracefully from the tail. Legs are reasonably long, free of any feathering and with four toes that are rather slim. Plumage is completely black but with a rich beetle-green sheen.

With the exception of the comb, which falls gracefully over the side of the head and should not obscure the vision, the general characteristics of the hen are the same as for the cock.

A well developed cock can reach weights of between 3 and 3.6 kg (7 or 8 lb) whereas the hen reaches between 2.2 and 2.7 kg (5 and 6 lb).

Sumatra Game (Plate 35)

Originally the Sumatra Game was known as the Black Pheasant, because of its appearance and plumage colour. In the early days of its history, when inhabiting its native Sumatra and other parts of Malaysia, there were probably six or even seven different strains, e.g. the Black, Blue, Gypsy-Faced, Pencilled and Spangled, all representative of the one breed.

Although different in outward appearance, they are in many ways allied to the Silk Fowl, or Silkie, insofar as they are black all the way through. They are black-boned, with black or dark blue flesh, very similar to that of the game birds such as the Pheasant or Partridge.

The Sumatra Game was first introduced into the British Isles in the early twentieth century by way of the USA, where it had already been established for many years and was, in fact, a very popular fowl and a showpiece of some notoriety. It had created a great deal of interest with the poultry fanciers, not only in the USA, but also with the rare breeds enthusiasts of the British Isles. It is certainly an unusual but nevertheless attractive-looking bird, with its large flowing tail and pea comb. However, it is not a bird to be kept for general purposes and allowed to run free-range, for, in wet cold conditions, its tail plumage soon becomes mud-spotted, damaged and unsightly.

137

Two of its main attractions are that it is a prolific egg-layer and, although being dark throughout, it lays white eggs. Secondly, the hens do make very good mothers, so much so that they will even sit on a clutch of pebbles, thinking that they are eggs.

Apart from the flowing tail, the cock has a pheasant-like appearance. The elongated body looks firm and muscular, with broad shoulders and a full and rounded breast. The wings are longer than average and quite powerful. The tail is long, flowing and quite majestic with its handsome sickle and covert feathers flowing gracefully as it struts along. The stance is firm and upright and the expression is bold and always alert.

The head appears somewhat small in relation to the body, with a strong, medium, slightly curved beak. Wide-eyed and always alert, this bird is very quick to notice any movement. The comb is of the pea-type and the ear lobes and wattles are surprisingly small for the size of fowl.

The neck is a fraction longer than average, strong and powerful with an abundance of hackle feathers, fitting neatly to the contours of the shoulders. The legs and feet are of average length, thick and powerful with strong muscular thighs befitting a game bird. Because of its firm stance, the thighs are set well apart. The four long and straight toes are well spread. The two spurs set just above the feet point backwards.

Being a light breed, weights are produced of approximately 1.6 kg (3½ lb) for the hen and 2 kg (4½ lb) for the cock, when fully mature.

The plumage is a rich black with a very prominent beetle-green sheen. Eyes are very dark ruby red with the face comb, ear lobes and wattles black or very dark ruby in colour. Legs and feet are a very dark olive shade.

Sussex (Plate 36)

The Sussex breed is an old English fowl, dating back to the early nineteenth century and was one of the few true breeds of poultry kept on farms for commercial purposes, because it was considered to be the ideal bird both for the table and egg-production. However, it was not until some years after its establishment as a domestic large fowl that it became established as a laying strain.

The Sussex is classed as a heavy breed of fowl and it lays slightly tinted eggs. It was the result of a cross between the Light Brahma and the Buff Cochin; eventually the Dorking also became numbered among its progenitors, together with the Old English Game Fowl. The end result was the Speckled Sussex. Since that time, as a result of the dedication and devotion of the ardent fancier, other strains have been developed, e.g. the Brown, Buff, Light, Silver and the Pure White. No doubt, as time passes, other strains will emerge.

Of all the strains, the Light Sussex is still recognised as the most popular of the breed and the Spangled is classed as the oldest member of the family. It was originally known as the 'Kent' Spangled Fowl and was of a totally different colour from those which we know today.

The cock is a very graceful bird, well proportioned and quite robust in appearance, with a solid stance. The body is rather broad with a flat back of reasonable length. Its breast is also broad, carried well forward, which gives the bird a proud appearance. The wings are neatly tucked in, tight to the body. The tail is of a reasonable length and lies at an angle of about 45° to the line of the back.

The skull is of medium size and seems to be rather small in relation to the size of the breed. The beak is short. The full bright eyes give an alert appearance. The upright comb is of medium size, evenly serrated and firm. The face is smooth and the ear lobes and wattles are of average size. The neck is of medium length with an abundance of fairly compact hackle feathers. The legs are rather short and stumpy and well spread apart, and the feet are quite strong and free of feathers. The four toes are straight, reasonably long and well spread out. A mature cock weighs between 4 and 5 kg (9 and 10 lb).

The general characteristics of the hen are similar to those of the cock, apart from the normal differences between the sexes.

The beak is white or horn-coloured. Eyes can be red, orange or brown, depending on the strain, but they are mostly red. The Light Sussex has orange eyes whereas the Brown Sussex may have either red or brown eyes. The comb, face, wattles and ear lobes are all red and the legs and feet are white.

Light Sussex
The Light Sussex has pure white feathering with the hackle feathers striped with black. The flights are also black but the remainder of the wing is pure white. The tail is black and the black feathers of the neck hackle should be completely surrounded by a white margin. As with the Rhode Island Red, the Light Sussex is still an important member of the poultry family and is widely used in cross-breeding by some of the commercial hatcheries in the production of the white-feathered

hybrids either for meat, when crossed with similar heavy breeds, or for the supply of white-shelled eggs.

Speckled Sussex

In the cock, the hackle is of a rich mahogany shade, each feather having a black stripe, tipped with white. The wing bows are speckled and the primary feathers are white with brown and black intermixed. The main tail feathers are white and black and the sickles are black with white tips. The remainder of the feathering is a rich mahogany shade, each feather having a small white spot at its tip. The speckled effect is the result of each feather being positioned in such a way as to show the three distinct colours. In the hen, the flight feathers are black, brown and white. The tail is black and brown with the same white tip. Hackle and body feathers are a rich dark mahogany and, in all other respects, resemble those of the cock.

Welsummer (Plate 38)

This breed of poultry is Dutch in origin and was named after the village of Welsum where it was first bred. It is said that the strain was produced purely by chance by a local farmer who kept several of the pure breeds of fowl. It was first imported into England in the early twentieth century and was, as far as one can tell, one of the newest breeds available at that time. The birds have practically the same colouring as the Barnevelder and the Black Leghorn and it is also very similar in shape to the Barnevelder. This is not surprising, as these two breeds figured strongly in the development of the Welsummer. Other strains used in its breeding were the Cochin, Wyandotte and also the Rhode Island Red.

The Welsummer has a single comb and red ear lobes. It is generally very active, ever scratching and searching to find those extra tasty morsels and yet always seeming to be quite contented. The Welsummer is quite an attractive bird in its own way. It had been some time since such a breed had been introduced and it soon became obvious that the docile Welsummer would become a very popular bird, not only for laying large brown eggs, which is still its main feature today, but also because of its non-sitting properties and usefulness as a table bird, due to its good body size.

The Welsummer cockerel is a very proud-looking bird that has a firm stance with an erect head and flowing tail feathers. It is truly a handsome bird in appearance. Alas, the lovely flocks of Welsummers which were once such a delight to behold are now a rarity, but one day, perhaps, because of their beauty, they may be revived.

Wyandotte (Plate 39)

The Wyandotte, of which there are many varieties, was first bred in the USA in about the early nineteenth century, in the State of Michigan. It was named after the picturesque town of Wyandotte, which is situated on the westernmost shores of Lake Erie, not far from the industrial city of Detroit. The first of the breed to be introduced into the Britisah Isles, in about 1880, was the Silver-Laced variety, which immediately created a great deal of interest with the poultry fanciers, as this was a new fowl with a most striking pattern to the plumage. However, since that time, as a result of the fanciers and geneticists experimenting in crossing the Wyandotte with other breeds, several other varieties have been introduced, e.g. the Barred, Black, Blue and Blue-Laced, Buff and Buff-Laced, Columbian, Gold-Laced, Partridge, Red and the White. Of all these, the author still regards the Silver-Laced as the most attractive of the Wyandotte breed.

When the Silver-Laced Wyandotte was first introduced into England, it was quite different in appearance to the birds of present-day standards, insofar as the colouring and lacing of the plumage was not quite up to the quality of the fowl which can be seen at exhibitions today. One of the other main points in which the birds differed was in the shape of the comb. In the early days, the strain had a rather large rose comb, similar to that of the Hamburgh breed; however, because of the experiments in cross-breeding over the decades, made by the more ardent fanciers, the modern variety now has a rose comb, which sets much closer to the skull and is described as a walnut-type rose comb.

After the introduction of the Silver-Laced Wyandotte came the Golden variety; in this the plumage is marked in the same way as the Silver, but the main colouring is of a rich golden brown rather than the silver which gives quite a contrast.

Although all these different strains have been created, the White Wyandotte can still be recognised as one of the most general purpose breeds, because of its utilitarian properties. Nevertheless, the breeding and development of the Wyandotte strain still remains the work of the fancier rather than the general poultry farmer, as breeding experiments to produce birds of the highest standard are still very much a scientific pursuit.

The general characteristics of the Wyandotte breed are practically the same the world over, i.e. the short head and broad shoulders. All varieties have yellow legs with, perhaps, the exception of the Partridge type, which sometimes has willow-shaded legs, as happens with the Black Leghorns.

The skull is rather short but broad. The beak is also short but well curved. The rose comb is firm and evenly set, square-fronted, tapering towards the back and following the curve of the neck. The face is quite smooth. Ear lobes are oblong and the wattles are of average length and well rounded. The neck should be of medium length and well covered with hackle feathers.

The rather short body is deep with a broad and well rounded breast with a straight keel. The back is also quite short with a rather broad saddle which gently rises to the tail. Wings are of medium size and well tucked in. The well developed tail is broad at the base and the main feathers are carried upright. The legs are of medium length, the thighs being well covered with an abundance of soft fluffy feathers. The four toes are straight and well spread apart.

The bird carries itself in a well balanced fashion and is always alert and quite graceful and, surprisingly, rather docile. The weight when fully mature should be approximately 3.6 kg (8 lb), certainly not less.

The cock and hen are very similar in appearance, apart from the normal differences between the sexes. Mature hens weigh approximately 2.7 kg (6 lb), but not less. They are, therefore, classed as a heavy breed. The eggs are tinted.

Yokohama (Plate 40)

The Yokohama, sometimes spelt Yocohama, is undoubtedly of Japanese origin, although it is not known by that name in Japan, where several varieties of the same breed exist, each having its own particular name, according to the area in which it originated.

The Onagodori Fowl, which is an ancient breed still in existence, is recognised as being the ancestor of the Yokohama. It is a fascinating creature with a long flowing tail and intelligent appearance. Its main feature is the long tail, which can grow as much as 900 mm (36 in) a year until fully developed. As the bird can live for fifteen years or more, its tail by then can be of considerable length. Because of its long tail, it spends most of its life on perches or high objects.

As the breed eventually found its way to Europe and, inevitably, was mated with other fowl, a new strain of the Onagodori developed. In the British Isles, this new strain became known as the Yokohama. In Germany, however, it became known as the Phoenix or Phoenix Fowl. The name 'Phoenix' is of European origin and has never been accepted in Britain. However, the Yokohama and the Phoenix are one and the same breed, the only difference being in the comb formation.

The Yokohama or Phoenix, as it was gradually developed in other countries, eventually lost its full length of tail because, when it was bred with cocks of other breeds, it was difficult to produce offspring in which the full flowing effect of the tail was maintained. Today the maximum length of tail found on the Yokohama in Britain and Europe is 900 mm (36 in) for a fully developed bird. Nevertheless, because of this unusual feature, it is necessary to keep these birds in specially designed pens for, if allowed to roam freely on wet and muddy ground, their tails soon become damaged and unsightly. Also, it they stand on their tails, the feathers may be pulled out. Therefore, to preserve an attractive plumage, the birds should always be allowed access to high perches at all times.

In the British Isles, there are two kinds of Yokohama: those with a walnut comb, i.e. the Red-Saddled and the White, and those with a single or pea comb, i.e. the Gold and Silver Duckwings and the self colours. It is the Red-Saddled single-comb strain that is of German origin and is known, in that country and throughout Europe, as the Phoenix. As the breed is being constantly improved with specialised breeding by dedicated fanciers, even the strains in the Duckwing colours are forming the walnut-type comb.

This is a very stylish fowl with intelligence and an air of regality. The type has a full and rounded breast with a long, deep body that tapers towards the tail. The wings are fairly long and closely tucked into the sides. Saddle and hackle feathers are quite long and complement the long tail feathers.

The head is rather small in relation to the size of the fowl. The beak is strong and slightly curved and the expression in the eyes is always alert. The comb varies according to the strain, i.e. a single, walnut or pea comb. It is usually the Red-Saddled British strain that has the walnut comb, especially for showing purposes. Ear lobes and wattles are oval in shape and rather small for the size of the bird. The neck is covered with an abundance of hackle feathers.

The breed is by no means a heavy one. The cock reaches weights of only 2.2-2.7 kg (5-6 lb) and the hen 1.4-1.8 kg (3-4 lb). The eggs are tinted.

The plumage of the Red-Saddled strain, which is one of the most popular, consists of red and white feathering. The breast and thighs have crimson or buff-red feathers and, in the cock, the back, saddle and wing bows are also crimson-red, hence the name.

Bantams

Although the general domestic poultry fowl has been in existence for thousands of years, the same cannot be said of bantams. Many bantams are Lilliputian variations of the standard domestic fowl, which have been bred for the exhibition arena but also with the aim of space saving. A select few bantams, however, are true breeds of dwarf domestic poultry e.g. the Belgian, Booted, Japanese, Nankin, Old Dutch, Pekin, Rosecomb, Sebright and Tuzo Bantams. Belgian Bantams include such varieties as the Barbu d'Anvers and the Barbu d'Uccle, two well known and most attractive fowl to say the least. In England and in Europe, the Booted Bantam is also known as the Sabelpoot.

Bantams have been known to exist for over 200 years, but very little is known about their history and origin. It was suggested in the early writings of the seventeenth century that the bantam fowl actually originated from the Indonesian and Oriental regions of the globe. However, this must have been pure speculation because, although the breed was at one time called the Java Fowl, no evidence has been found to substantiate such a theory. On the other hand, neither has the theory been categorically denied, for the Black Rosecomb Bantam that we know today is also known as the Black Java in Europe. As it was general practice to name the poultry breeds after their place of origin, it is possibly a correct assumption that they originated from the East.

Aldrovandi in the sixteenth century, Bonington Moubray in 1842, W.B. Tegetmeirer, Lewis Wright, L.C. Verrey in 1893, W.F. Entwistle in 1899 and Harrison Weir in 1902, all authors who were considered to be the foremost poultry writers of their day, wrote about 'the pigmies of the larger poultry breeds', which tends to substantiate the belief that the bantam fowl was well and truly established even in those days. Since that time, the development of the species has been steadily progressing, so much so, that nowadays thousands of bantam fowl exist, whereas, a hundred or so years ago, only a few adorned the countryside of England.

145

This diminutive but picturesque little fowl has many relatives and a history that dates back to the late sixteenth century, when Ben Johnson and William Shakespeare were actively engaged in the organisation of the Globe Theatre in Southwark in the City of London. It is believed that the Belgian Bantam, first bred in Antwerp, was recorded by some old Dutch writers in about 1610 but, at that time, the breed was known by another name. In earlier times, it existed mainly on the Continent, where Belgium, Holland and Germany were considered to be the major areas of poultry development. It was not until 1910 that the Belgian Bantam was brought to Britain and, in 1911, it made its debut at the poultry exhibition at the Crystal Palace, which was THE major poultry event in those days.

There are two distinct types of Belgian Bantam: the Barbu d'Anvers, which has a rose comb and shanks that are completely devoid of feathers, and the Barbu d'Uccle, which has a single comb and an abundance of feathering on the legs. Both types have a similar beard and because of this they are frequently referred to collectively as the Belgian Bearded Bantam (barbu = beard). A feature noticeable in both is that, when viewed from the front, they tend to have an owl-like appearance, because of their wide eyes, short slightly hooked beak and their whiskers.

The Barbu d'Anvers, the first of the breed, has a very long history and the Barbu d'Uccle is the result of Man's genetical expertise in crossing the Booted Bantam (or Sabelpoot) with the Barbu d'Anvers; within the short space of three years, after further in-breeding, there emerged the truly delightful Barbu d'Uccle.

During the course of time, other strains have been developed so that there are now five different types of Barbu Bantams. However, it is only the Barbu d'Anvers and the Barbu d'Uccle that have been standardised in the British Isles. Perhaps in the not too distant future, the third type to be developed in recent years, the Crested Barbu Bantam, may be accepted into the British Standards list. In addition to these parent breeds, many new varieties now exist which have plumage of varying colours and designs, every one of them a beauty, and anyone who admires the breed will see a good deal of character in their features.

In Europe, the breed is still extremely popular and its survival is undoubtedly assured because of the dedication of those fanciers who are members not only of the Belgian Bantam Club in Belgium itself but also of the American and British Belgian Bantam Clubs.

Barbu d'Anvers

The cock stands very erect, almost upright, with its head held high and its prominently curved chest pushed well out. A very proud-looking bird, it has an abundance of well developed neck hackle that extends almost to the saddle. This is not surprising, as it has a very short back that slopes acutely downwards to the tail. The tail, which is carried well up, has only two main sickle feathers and, because of their short length, they do not have the full curve, as in most other breeds, but are sword-like in shape. The remaining tail feathers lie in a fan shape and are well covered at the base by the long saddle hackle. Because of the bird's upright stance, the wings lie almost perpendicular, with the primary feathers practically touching the ground.

The head is of average size but seems larger because of its beard. The short beak curves slightly at the end and the colour blends with the plumage. The rose comb is very broad at the front and slightly hollow. A small number of ridges or spikes adorns the middle, which then tapers to a leader at the back. The large prominent eyes are dark in colour and most of the face is covered with small feathers. The ear lobes are very tiny and the wattles are almost non-existent.

As far as the legs and feet are concerned, the thighs are short in relation to the shanks, which are of average bantam length and completely devoid of feathers. The four toes have nails of the same colour as the beak, i.e. light at the tip fading to a deeper shade at the base, again blending with the plumage colour.

The hen has a well rounded or arched neck with a fully developed neck hackle that is more compact around the neck than in the male. The tail is much shorter than that of the cock and not quite so compact. Otherwise the general characteristics are similar.

Barbu d'Uccle

This breed is very similar to the Barbu d'Anvers but with the following exceptions. There is a single comb which is evenly serrated and an abundance of leg-feathering, extending down the full length of the shanks and culminating in foot-feathering. The cock is slightly taller than its ancestor, because of its extra leg length, although this is not readily noticeable because of the leg-feathering. Apart from the natural sex differences, and the tail being more fan-shaped and devoid of sickle feathers, the hen is very similar to the cock.

147

Colour Varieties
Of the many colour variations developed over the years, several have achieved as near perfection as possible and the majority have now been given British Standards descriptions. These are: the Black, Black-Mottled, Quail and Blue Quail, Cuckoo, Lavender (Figure 31), Millefleur, Porcelaine and the White. Every one of them is a joy to behold and, when all are exhibited in one show, as often happens, you can stand for long periods just admiring their beautiful colours and intricate designs.

The Black has an entirely black plumage with a beetle-green sheen whereas in the Cuckoo each of the feathers is barred with bands of dark grey on a lighter ground. In the Quail, the head feathers in the male are black with a beetle-green sheen and are laced with gold. The beard is a rich golden buff, blending upwards to a darker shade. The

Figure 31. Barbu d'Anvers Lavender Quail hen.

neck and saddle hackles are a rich black laced with golden buff, the saddle hackle being slightly darker in shade. The wing bows are a light golden colour with the lower half of each feather being distinctly black. The wing bars are of a yellow buff colour and the secondaries are partly buff and partly black. Primaries and tail feathers are completely black with a beetle-green sheen mainly in the tail feathering. The side hangers are also black but laced with chamois. The breast and thighs are of an ochre colour and the remaining underparts are a greyish brown, each feather having a gold tip. Legs and feet are slate-grey.

The Blue Quail is very similar to the Quail, the main difference being that the black is replaced by a shade of pale grey-blue.

The Millefleur is truly remarkable for its intricate colour patterns and designs. There are four main colours: black, chamois, red and white. The head feathers are reddish orange with white tips and the eyes are orange with black pupils. The ear lobes and wattles are red but almost non-existent. The beard consists of black feathers laced with chamois and sometimes tipped with white. The neck hackle is black with golden shafts, each feather having a black stripe at the end and tipped with white. In both sexes, the back is orange-red, fading slightly towards the saddle hackle. Wing bows are a deeper red, each feather having a white tip. The wing bars are also red but have a gentle brown or russet shade with black spots towards the ends of the feathers which are tipped with white. The secondaries, like the primaries, are black but with the lower part of a chamois colour. The remaining wing feathers are chamois with a black bar or spot and end in white tips. The black tail feathers, with their beetle-green sheen, are edged with chamois and tipped with white. The rest of the plumage is a golden-chamois with black markings and tipped with white at the ends. Legs and feet are slate-blue.

The pattern of the Porcelaine is similar to that of the Millefleur except that the black shading is replaced by pale blue and the undercolouring is wheaten or light straw.

Booted Bantam

This exquisite but quaint little fowl which has, unfortunately, been pushed into historical obscurity, is almost forgotten as a 'show' fowl in favour of the more beautifully coloured Barbu d'Uccle. It is one of the oldest known breeds of true bantam and can be traced back as far as the early nineteenth century, but little is known about it before then. It most probably originated in Europe as it resembles very closely the Barbu d'Uccle. Belgium or Holland are, therefore, most likely the countries in which it was first discovered. The main difference between the Booted Bantam and the Belgian Barbu d'Uccle is that there is no muffling or beard around the chin and a rather larger pair of wattles than usual for this size of fowl. In the early years of the nineteenth century, poultry writers advocated that, 'they had good muffs and whiskers' however, as the Booted variety has been developed over the years, it has gradually lost its neck muffles.

Of all the bantam breeds, the Booted Bantam, or Sabelpoot as it is sometimes still called, is considered by the majority of bantam fanciers to be the most prolific egg-layer but, alas, the eggs are not much larger than those of a pigeon! Also, the hens have the strong mothering instinct which makes them good sitters. Again, it is a breed of bantam that has no counterpart in the standard breeds.

They are delightful little creatures; the White is not much larger than a good-sized snowball and is perhaps the tamest and hardiest of all the true breeds of bantam. Like the Pekin, it is an ideal pet for children and can be left to roam the garden at will, without damaging the flowers and vegetables (providing of course that it is fed at the appropriate times with the correct balanced feed). It should not be given free-range if other domestic pets, such as cats and dogs, are at liberty in case of attack. The only problem with the pure white variety is that it soon becomes dirty and unsightly if allowed the freedom of the garden, especially during inclement weather; even if allowed to remain in strong sunlight for a few days, the pure white plumage tends to take on a yellowish tinge in places, just as it did in one special pure white fowl recently bred by the Author. This yellowing of the plumage tends to spoil the appearance somewhat and would certainly lose points if birds are being exhibited.

Colour Varieties

There are four main colour varieties of Booted Bantam: the Black, White, Black-Mottled and the Millefleur. The Millefleur variety has a most attractive pattern of intricately interwoven and blending colours of black, chamois, orange and white with a tinge of green on the tips of some feathers. The Black is as pure black as possible with a lustrous sheen and the Black-Mottled has a beetle-green sheen with white tips to each feather.

The Booted Bantam is somewhat smaller than most other bantams, which makes it quite amusing at times. ('Booted' is another term for the leg-feathering growing from the shanks and toes of poultry, as in the Cochin, Brahma and Pekin fowls.)

Japanese Bantam

The birthplace of this dumpy little bantam lies deep in the farming communities of Japan and its history probably dates back far beyond the nineteenth century, although little evidence can be found to substantiate any development before that time. It was not until the beginning of the seventeenth century, when trading began to improve between Europe and the Orient, that the first Japanese Bantam was introduced into Britain. Since those early days, many new strains have been introduced, all of which have the same dumpy appearance. Not only were new varieties imported into Britain from the Japanese mainland in years gone by, but the development of the breed in Britain made enormous strides during past decades. This was due to the dedication of the bantam fanciers, who are responsible for the dozen or more variations that exist today, e.g. the Black, Blue, Buff, Cuckoo, Grey, Lavender, Mottled, Red, White and most other combinations of colours of the Old English Game plumage, as well as the Black-Tailed White, (Figure 32), Black-Tailed Buff, Columbian etc.

In spite of all the experiments that have taken place, this *petite poulet* still has a rather odd, pugnacious and grotesque look when compared with other breeds of bantams, as if completely deformed; yet, to the eyes of the fancier, it is a beautiful bird. It is reasonably quiet and certainly not as quarrelsome as other bantams. It is said by many to be an antiquated little fowl and rather quaint, with its small chubby-looking body and extremely short legs. The newcomer to the world of poultry, seeing it for the first time, might well think that it is always squatting but it is the short, stubby, almost non-existent legs that give it this appearance. Because of their short hocks, breeding proves very difficult. The main drawback is the lack of leg length, which makes it rather difficult, to say the least, for the cock to position himself correctly to mate with the hen, thus resulting in poor egg fertility. Dedication and perseverance seems to be the only answer to producing specimens of good quality and appearance.

Every care should be taken when breeding the 'Jap', as it is sometimes affectionately called, for they are rather delicate and cannot withstand extremes of temperature. Cold damp conditions should, therefore, always be avoided as far as possible. Because of

151

Figure 32. Japanese Black-Tailed White cock.

their delicate nature they require extra special attention from the breeder if he intends to rear them to a high degree of perfection or to exhibition standards. In view of their delicate constitutions, it is best to avoid early hatching in January and February for obvious reasons. It is much more advantageous to wait until the warmer months of June, or even July, so that the chicks can benefit from the warm summer months and still be fully grown by the time winter sets in.

Like most other bantams, they are quite reasonable layers and certainly good mothers; they will nurture their little offspring for as long as is necessary. They are certainly an indoor fowl as, because of their short leg length, if they are allowed free-range, their plumage soon becomes damaged and unsightly.

The tail feathers of the cock stand remarkably upright, which gives it a most unusual sabre-like appearance; the leaders or main sickle feathers are very long, slightly curved and gently leaning over the base of the back. The head is quite broad and is topped by a very large comb that is rather long for the size of fowl, although it does give it a balanced look in relation to the size of its tail and is much larger than that of the female. The comb has five serrations continuing along the full length of the head and ending in a long curve at its base. The eyes are full and wide and the beak is strong and stubby, while the wattles are well balanced and also very large. The body is short but quite broad. Although the neck is very short, the neck hackle feathers, like those of the saddle hackle which drape gently over the body, are exceptionally long. The wings of both male and female are also long for the size of body, to such an extent that they almost touch the ground. The full breast is carried well forward, again to give the body balance. Its legs are almost non-existent, with the feet being just visible from under the plumage.

The hen has characteristics that are very similar to those of the cock, except for the natural female attributes. The tail is well spread with a uniform feather formation.

The profusion of colours and variations in the numerous strains in existence today, are a sight to behold, but, regrettably, are too numerous to mention.

Nankin Bantam

Although the Nankin (Figure 33) is still accepted as an exhibition bantam, it has never really created much enthusiasm among the ardent fanciers, probably because of its dark blue legs and also because of its size, for it is the largest of the bantam breeds, neither standard nor bantam-sized. Since before the 1850s, the Nankin has been classed as an exhibition breed, but, alas, has achieved no accolades in the world of bantams; yet it can be quite attractive with its rich orange- and chestnut-coloured plumage — the base colour being very much like that of the Buff Cochin. The side hangers and minor sickle feathers are a dark buff or brown, gradually deepening to black in the main sickle feathers.

It is one of the most ancient of all the bantams and originates from the Asian continent, most probably from eastern China. Early writings indicate that there is a strong possibility that the Nankin was

Figure 33. Buff Nankin Bantam cock.

among the breeds of poultry introduced in years gone by and, as a result of crossing with other fowls, could well be one of the ancestors of the Buff Cochin because of its similar appearance. The Nankin Bantam is a true breed with no counterpart in the larger breeds. It is a rare breed of bantam and, apart from the Rare Breeds Society under the auspices of the Poultry Club of Great Britain, there is no specialist fanciers' club in existence to further the development of this quaint fowl.

The Nankin Bantam is recognised as being a first-class layer, equal to any farmyard fowl. The hens are good mothers too and also make very good foster parents; a hen which is sitting will readily accept any day-old chicks and will nurture them through their early days.

To many, the Nankin Bantam can be individually attractive yet it is not a fowl that takes a prominent place; nevertheless, its merit has been recognised by the dedicated few who have persevered with its breeding over many years.

Like the Japanese Bantam, the Nankin cock has long wings that almost sweep the ground and a tail that is carried very high and well spread out. As the breed has been gradually developed over the years, many are now being produced with rose combs and these are slowly but surely being more widely accepted. First impressions are that it has a rather conceited look; nevertheless, it has a sprightly gait, which at times can be quite amusing.

Old Dutch Bantam

At some stage during the late 1960s, the Customs and Excise authorities in Britain must have been somewhat negligent in their duties, for it is believed that this ancient Dutch breed of pure bantam was brought to the shores of Britain in the form of fertilised eggs, a practice which was, and still is, illegal. It was not until 1970 that the breed was legitimately introduced to British soil with the importation of live birds from the Netherlands.

Although the development of the Old Dutch (Figure 34) has continued within this country, it is still a very rare breed of bantam and one that is not readily accepted by the majority of the poultry fraternity, probably because of its peculiar shape and small size. It has, however, been given a place in the world of exhibition poultry, although it is still in the hands of only a few breeders. It is these fanciers who have produced some remarkable results with the development of the Blue, Cuckoo, Partridge, Blue-Laced and the Blue, Gold and Silver Duckwings, all of which are true bantams. The one colour that seems to be exceptionally difficult to produce, and which has eluded the experts for many years, is pure black. No doubt the geneticists will, in time, achieve the perfect strain.

When fully grown, the hen bird weighs about 500 gm (18 oz) and is not much larger than an outsized pigeon. Its tiny eggs are tinted in colour and are most useful for culinary purposes.

Pekin Bantam

The Pekin Bantam has descended from a very long line of true

Figure 34. Old Dutch Bantam cock.

miniature fowl and has no counterpart in the standard breeds of poultry. It is, however, so very similar in appearance to the Cochin, another Asiatic breed, as to suggest some mixing of the genes during their development. This is a theory that will probably never be proved yet the Chinese and other Oriental races are well known for their interest and expertise in dwarfing birds, as well as plants (bonsai), and it is possible that the Pekin may be a result of their experiments. On the other hand, there are specialist breeders today who will confirm that the Pekin breeds absolutely true to type, which confirms that it has been in existence for a very long time.

It is believed that the 'Lilliputian Pekin' was first discovered in the Pekin district of China in about 1860, during the Chinese war. A young army officer who had taken part in the fighting during the Anglo-French expedition, discovered these tiny fowl in the summer palace of the Emperor of China, which had been over-run and ransacked by the marauding hordes. The Emperor was interested in many of nature's beautiful things, including miniature fowl, such as the Pekin Bantam.

In the mid-eighteenth century, travellers continued to trade with the Far East and those who had spent some time in the area of Pekin, completing their trading arrangements, also discovered these miniature fowl; it was with their assistance that the breed eventually found its way to the shores of England in about 1865. Like most other breeds of poultry, it was named after its place of origin.

As new strains were gradually developed over the years, the Pekin Bantam was discovered to be one of the easiest of all bantam fowl to rear. It is a charming little fowl by any standards and, somehow the Chinese have been able to impart a sense of character into this lovely creature.

Colour Varieties
Since the early days, and particularly in more recent years, several varieties of the Pekin have been developed. Apart from the Black and the White, there is the Blue, which has a rich pale blue plumage; the Buff, which is evenly shaded throughout; the Cuckoo, with a regular barring of dark slate on a lighter grey background; the Mottled which is evenly mottled with white tips at the end of each black feather and an overall beetle-green sheen and the Barred, whose white feathers each have black bars at regular intervals and end with a black tip. The Columbian has white plumage with black markings along the middle of each feather and a black edge all the way round. The tail feathers are black and beetle-green sheen and the primary and secondary feathers are almost completely black. The Lavender, despite its name, is not necessarily true to colour, as it appears to have a silvery blue tint in its plumage. In the Partridge cock, each feather, like that of the Columbian, has a distinct black stripe down the centre of each feather with a black tip and fringe. The head and neck hackle are orange or golden-red, fading slightly towards the shoulders and the saddle hackle has similar shading. The remainder of the plumage is a rich green-black but the wing bows are a rich crimson. In the hen, the head and neck hackle are much lighter than in the male, being more of a straw or light golden colour but with each feather having the dark black stripe down the centre. The remainder of the plumage is a light partridge brown, i.e. evenly pencilled with a darker shade of greenish black. The healthy birds having an attractively fine sheen on the plumage. The eggs are creamy white.

Rose Comb Bantam

The history of this classic breed is still obscure but it is certainly considered by most people to be of ancient descent. The literature confirms that this true breed, in its present form, did not exist during the Shakespearian era or even in the late sixteenth century. There is, however, the possibility that it emerged in the early eighteenth century, probably as a result of experiments in cross-breeding.

The general consensus of opinion seems to be that the first of the Black Rose Comb Bantams were probably the descendants, either directly or indirectly, of the Black Hamburghs. Nevertheless, no matter how they materialised, they are a true breed of bantam.

Of the three different varieties of this bantam in existence within Great Britain, i.e. the Black Rose Comb (Figure 35), the Blue and the White, it is the Black Rose Comb, the first of the breed, that is most popular. This is confirmed by the number of breeders who continually enter this particular strain in the extensive number of

Figure 35. Black Rose Comb cock.

exhibitions that are held within the British Isles. The breed in Europe is known as the Java Rose Comb and can be found in many colour variations; this could perhaps be a pointer to the birthplace of the original Rose Comb Bantam.

They were at one time one of the most prolific egg-producers of the bantam breeds, although they were not much good as mothers. If a hen is mated with a good Black Rose Comb Bantam cock, the progeny will invariably breed true to type, developing into good-quality show birds, capable of maintaining the high standards set by their ancestors.

The body of the Rose Comb Bantam is short and slightly broad, with symmetrical curves from head to tail. The breast is held well up and the wings are of true bantam shape and carried low — a particular feature of the classic 'Lilliputian' breeds. Its main feature is, of course, the rose comb that gives the breed its name. The comb is well filled with bobbles or spikes at the front and is set firmly and square on the head, tapering to the back with a leader. The comb lies straight and in line with the body, tapering to a fine point and gently rising from the front of the head to the rear. The ear lobes are white.

Sebright Bantam

The majority of people who know anything about agriculture will have some knowledge of Sir John Sebright, not only for his skill as a breeder of Shorthorn cattle and other animals, but also for his skilful handling of the development of this diminutive and most attractive little fowl.

It was in the early eighteenth century that the first of this true bantam breed came into existence, as a result of Sir John's experiments on in-breeding with a standard bantam and a Polish fowl. As his experiments continued, with the aid of a miniature Golden Hamburgh, and later with a White cock, there eventually emerged the Gold- and Silver-Laced bantams. These were eventually named after him in recognition of his achievements in producing this most beautiful breed of bantam. It is one of the oldest of the British varieties of bantam and even now, there are still only two colour variations available: the Gold-Laced (Figure 36) and the Silver-Laced. Today, they continue to take pride of place at many of the poultry exhibitions.

Within a very short time of their creation, they were arousing so

159

Figure 36. Golden Sebright cock and hen.

much interest with other poultry breeders that Sir John, himself, in about 1810, formed the Sebright Bantam Club, which is still in existence today, having grown from strength to strength over the years with an enthusiastic membership of dedicated fanciers and breeders.

Alas, their laying qualities leave something to be desired for, as a general rule, they will only lay eggs during the spring and early summer months, for the sole purpose of perpetuating their species. Even now there is great difficulty in breeding Sebright Bantams, because the degree of in-breeding over the years has resulted in many cocks being sterile. Even fertile eggs do not always produce chicks which are true to colour. As they are small fowl, weighing just over 0.45 kg (1 lb) when fully grown, they are by no means suitable as table birds and are therefore bred purely for their charm and beauty.

Unlike most other breeds, the Sebright seems to walk on tiptoe all the time and has a strutting gait, with its head held high, giving it a self-important appearance which is most unusual. It has a well-developed rose comb and, in years past, face and wattles were usually of a dark purple colour. Nowadays, however, the head colour seems to be a deep shade of red; in fact, it is not uncommon for exhibition birds to be perfectly red. Unlike most other male birds, the Sebright cock is similar in appearance to the hen, the feathering being very much alike and the neck and saddle feathers matching perfectly. The

160

tail also is exactly the same and, in both male and female, the wings are of the true bantam type and, as usual, carried very low. The tail is well spread and in line with the head and back and is completely devoid of sickle feathers. The prominent and well rounded breast is held proudly forward. Feathers are quite uniform in width and, in a good specimen, the tips are always rounded and not pointed as in most other breeds.

The plumage of the Golden-Laced is a golden chestnut colour, each feather laced with a black edging that has a slight beetle-green sheen.

The Silver-Laced has plumage of a rich silver-white colour, again laced with a black edging and, as with the Golden-Laced, the markings are identical in both cock and hen.

In both male and female, the eyes are pure black. Comb, face, wattles and ear lobes are a deep purple or dark claret and the legs and feet are a slate-blue.

This is a true bantam, originating in Great Britain, that lays creamy white eggs.

Tuzo

The Tuzo (Figure 37) is a recent addition to the poultry world of Great Britain, so new in fact that it was as recently as 1971 that the first of the breed was accepted by a few dedicated fanciers, who were interested not only in preserving the species, but in introducing new blood into the field of poultry development. Being characteristically very similar to the Asil and, in plumage colour, to the Old English Game, it soon became accepted as a bird worthy of exhibition. Consequently, it has now been approved and given British Standards status by the Governing Body of The Poultry Club of Great Britain. Despite these similarities, it is a true bantam for, like other pure bantams, it is known to breed true to type. Being an ancient breed from the Far East, it is possible that its early ancestors may have inter-bred with Asil fowl in the wild; however, this has not been established.

Although this particular strain is new to the British Isles, in Japan, its native country, several varieties of the Tuzo adorn the farmyards. No doubt, in Britain, as time progresses, many other varieties will also emerge, either because of imports of the breed or as a result of genetic wizardry by the experts. It is a sad fact, however, that the breeders of this new found fowl are very small in number.

Figure 37. Tuzo cock.

Like the Asil, the Tuzo stands quite upright, with a slightly curved neck, but, unlike the Asil, it is not particularly designed for fighting. Its wings are surprisingly short, unlike most other true bantams which have wings nearly touching the ground. The tail is almost horizontal to the body and the overall plumage consists of rather short feathers, quite firm and rather coarse to the touch. The beak is short and stubby, slightly curved or hooked, and, at the forehead, sits the tiny comb that has a triple row of spikes or studs.

Turkeys

The turkey is a very ancient member of the poultry family and its history dates back many thousands of years, to before the birth of Jesus Christ. Fossil remains found in Central America confirm that North America and the Central Caribbean region were the original 'Land of the Turkey'. The Aztecs, Incas and Mexican Indians also figure strongly in its history, for many of their paintings depict the turkey in different postures, each having a particular, usually religious, meaning. Turkeys played an important role in the ancient Aztec and Mexican ways of life; they were symbols of the powers of goodness and, because they were held in such high esteem, were often sacrificed and offered up to their gods during ancient rituals.

Although there is no evidence to indicate that these ancient civilisations kept turkeys on a domesticated scale, it is reasonable to assume that they cared for them in some way because many of their ceremonial robes, especially their headdresses, consisted of most unusual, long, attractively coloured feathers. To design and maintain this colourful attire, they must have had access to the turkey, or a similar fowl, to obtain the feathers, which were often stained with natural dyes. However, the Peruvian and Mexican Indians were interested in the turkey not only for its decorative plumage but also for its succulent white flesh.

As time progressed, there is no doubt that the turkey fowl became more and more popular with many of the tribes that inhabited Central America, for it is recorded in American history that the nobility of Mexico and neighbouring States kept large flocks of similar birds, amongst which turkeys favoured most strongly.

It was later discovered that the wild turkey inhabited not only the countries of Central America, but also the North American continent, as far north in fact as the Canadian border. This was confirmed by the Pilgrim Fathers and their sixty-seven followers who, on 6 September 1620, sailed from Plymouth, England, in the good ship *Mayflower* destined for North America. They took with them not only goats and pigs, but also live chickens to supply fresh eggs and meat. By 1622, the emigrants had overcome their initial hardships and had established friendly relationships with the North American Indians. It was then that the Pilgrims learned of the large wild birds that roamed the forests and marshlands of the New World. To their joy and amazement, they soon discovered that these large wild birds could supply them with an abundance of fresh tender white meat. As a result of this newfound wealth, which in those days was a guarantee of survival, the turkey is now the traditional festive meat used at the Thanksgiving Celebrations each October in the USA.

Spain also played an important role in the history of the turkey, but it was not introduced into the Iberian Peninsula, where it later became domesticated, until the late fifteenth and early sixteenth centuries when the Spanish explorers discovered Mexico. Even today, in those American States bordering the Caribbean and in the Everglades of Florida, the wild turkey is still in existence, in its natural state. Being a bird of the forests, it usually haunts the most densely wooded areas but it is also found on the swampy marshlands living in flocks, each cock having numerous wives. Because the cock frequently eats the eggs, the hens conceal their nests very carefully. Each hen lays about 10-16 eggs during the laying season.

Although the modern turkey can supply both meat and eggs — like the chicken — it has never aroused the same keen interest with the general domestic poultry keeper, possibly because of its size and the extra attention needed to keep it strong and healthy. The turkey fowl is not considered to have a robust constitution, and it is certainly not as intelligent as the domestic hen, yet its make-up and feeding pattern are very similar. The main difference is that the turkey is now bred principally for meat-production and those birds that are reared for egg-laying are reared solely with the reproduction of the species in mind. Nevertheless, there are signs that the turkey is now becoming accepted as a domestic fowl partly as a result of the recent advances made as a result of careful selection and breeding by the turkey hatcheries, and also because of the development and availability of high quality feeds.

The first imports of the breed into the British Isles occurred just prior to, and during, the reign of Queen Victoria in the 1820s and 1830s. Shortly afterwards, the fowl was acclaimed as a delicacy and was readily accepted. Surprisingly, the only major development of the breed to take place since that time has been in the world of commercial poultry. Turkey meat-production on a commercial scale has expanded so much that, today, we take the turkey for granted, as a regular and cheap source of protein. However, turkeys are still held in high esteem and are eaten by many households each Christmas.

Brush Turkey

Many decades ago, when colonisation was occurring in the southern hemisphere, a bird very similar to the turkey was discovered on the continent of Australia. This bird, called the Brush Turkey because it lived in the brushlands, is in no way related to the turkey discovered in the northern hemisphere. It belongs to a group of birds called *megapodes* — game-birds that lay their eggs in mounds which they construct specially for the purpose. Although the cock is small, it bears a striking resemblance to the domestic turkey, with well defined white markings on its shiny black plumage, and a red head and neck devoid of feathers. It has yellow wattles, which during the breeding season, grow much bigger than those of the ordinary domesticated bird. The hen has no wattles whatsoever. Like most other wild turkeys, their anatomy does not permit sustained flight and so they cannot travel vast distances. Nevertheless, they are able to fly up to the branches of trees where they roost whenever possible.

During the breeding season, the cock Brush Turkey and, on occasions, the hen bird as well, scratch together a mound of leaves, grass and sometimes twigs that can reach heights of 180 cm (6 ft) or more. A short time after its completion, the mound begins to heat up as a result of decomposition and fermentation. The hen bird then makes a hole near the top of the mound in which she lays an egg, filling in the hole afterwards. Over a period of about 10-12 days, this operation is repeated until about 8 or 10 eggs have been laid, arranged in a circular pattern around the mound, each in its own hole. After about 56 days, the chicks break free from their shells, fully covered in plumage, and, even at this early stage, are capable of finding their own food; some can even fly after a few hours in their new environment. However, they invariably stay close to the mound where they have the protection of their parents.

White Beltsville, Blue and Buff

Since the turkey was first discovered, some development in the domestic field, although not extensive, has taken place which has resulted in new strains being produced. For instance, there is the White Beltsville, developed in the USA, which can reach weights of up to 8.2 kg (18 lb) or so, and the Blue and the Buff (Figure 38) turkeys, also of American origin, that have very similar characteristics and are much heavier birds than the White Beltsville, reaching weights of 10.4 kg (23 lb) and sometimes more.

British White

The British White is considered to be the most popular of the commercial turkeys, partly because it is quick to mature and also because of the large quantity of breast meat which it can produce (it can reach weights of over 15.8 kg — 35 lb). The British White, as it is known throughout the British Isles, is not the only strain of white turkey available for white turkeys have also been produced in some European countries for a long time; its Continental equivalent is known as the Holland White.

Cröllwitzer

Another strain which is of European origin is the Cröllwitzer. This is a German breed considered to be one of the most attractive breeds of the turkey family because of its diagonal black stripes tinged with silver on a pure white plumage. It has been known to reach weights of 16.3 kg (36 lb) and over.

Norfolk Black

Within the British Isles, the two most important strains that exist are the British White and the Norfolk Black; the latter was developed in the county after which it was named. It is considered to be the turkey for the gourmet because of the fine texture and delicate flavour of its flesh. It was the first of the domesticated turkeys to be bred in the British Isles and is still popular today. It is considered to be the Blue Riband of the turkey world, although it is the British White which is more popular in the commercial world because of the heavier weights which it can reach. The general appearance of the British White and the Norfolk Black are very similar, the only differences being in the colour of the plumage. The White, however, produces a much greater

Figure 38. Buff stag turkey.

Figure 39. Mammoth Bronze stag turkey.

167

quantity of breast meat than the Black, perhaps at the expense of flavour. Face, throat, wattle (note the singular) and caruncle are a rich bright red in both breeds. The cock birds have a wattle that hangs over the beak and a conspicuously long tuft of hairs on the breast.

Bronze (Mammoth Bronze) and American Mammoth Bronze

Other breeds include the Bronze, or Mammoth Bronze (Figure 39) as it is now generally known. Although the Mammoth Bronze was developed in Britain, its American cousin, the American Bronze, also figures strongly in its make-up, for the two strains have been inter-bred to produce the now well known American Mammoth Bronze. In spite of its name, it can reach weights of only 12.7 kg (28 lb) or so.

Appendices

APPENDIX 1 Commercial Hatcheries, Breeders and Suppliers of Pure Breeds of Poultry

For the newcomer interested in keeping the pure breeds of fowl, commercial hatcheries specialising in such pedigree stock are listed below. Apart from these commercial breeders, there are also many breed clubs in existence whose members specialise in rearing not only the pure strains but also the rare breeds of poultry, e.g. the Old English Game, Asil and Malay. These members are the dedicated fanciers who strive to prevent the pedigree breeds from becoming extinct and they do sell their surplus stock from time to time. Further information can be obtained from the Poultry Club of Great Britain, 24 Faris Barn Drive, Woodham, Weybridge, Surrey KT15 3DZ.

United Kingdom

Cyril Bason (Stokesay) Ltd, Bank House, Corvedale Road, Craven Arms, Shropshire SY7 9AN.
D.E. Bonnett, Longcroft Farm, Ramsden Heath, Billericay, Essex CM11 1JN.
R. Bows, Home Farm, Laughton, Gainsborough, Lincolnshire DN21 3PS.
C.E.F. Agricultural (Holdings) U.K., 2 Pinfold Lane, Scarisbrick, Ormskirk, Lancashire L40 8HR.
S. Davies, Stonehill Farm, Blagdon Road, Collaton St Mary, Paignton, Devon.
S. Earl, Holm Lea, Station Road, Salhouse, Norwich, Norfolk.
Hamers Chicks & Incubators, Bradshaw, Bolton, Lancashire BL2 4JP.
R. Jenkins, 27 Station Road, North Walsham, Norfolk.
Johnsons of Mountnorris, 11 Porthill Road, Mountnorris, Armagh BT60 2TY.

D. MacLeod, No. 7 Lochend Barrock, Castletown, Thurso, Caithness KW14 8TA.
Novoli Poultry Breeding Farm, The Street, Purleigh, Essex.
O'Kane Hatcheries Ltd, 117 Race View Road, Broughshane, Co. Antrim BT42 4JL.
O'Kane Poultry Ltd, 170 Larne Road, Ballymena.
Pennine Poultry Farm, 234 Carterknowle Road, Sheffield, South Yorkshire S7 2EB.
G. Potter & Sons, Howefield, Baldersby, Thirsk, North Yorkshire YO7 4PZ.
Southern Pullet Rearers, Greenfields Farm, Fontwell Avenue, Eastergate, Chichester, Sussex.
Sunnyside Poultry Breeders, Blockett Lane, Little Haven, Haverfordwest, Dyfed SA62 3UH.
J. Thom, 22 Townsend Terrace, Kintore, Inverurie, Aberdeenshire.
Wrights Chicks Ltd R.A., 21 Ballinderry Road, Lisburn, Co. Antrim BT28 2RP.

APPENDIX 2 Commercial Hatcheries, Breeders and Suppliers of Hybrid Poultry

---

For the domestic poultry keeper who wishes to purchase stock for meat- or egg-production, the following list of commercial breeders may be useful. However, it should be mentioned that, if a special shipment of stock is required, these major breeders usually prefer to deal in large quantities. Nevertheless, providing that the purchaser is prepared to collect the stock himself, then these breeders will invariably supply poultry in smaller numbers of a dozen or so.

There are also many other reliable breeders who advertise in the classified sections of local and country newspapers and magazines and supply point-of-lay poultry. These smaller breeders may be more accessible and suitable for the small-scale poultry keeper.

Australia

Hygenic-Lily Ltd, P.O. Box 11, Alexandria, New South Wales 2015.

Irish Republic

Arbor Acres Ireland Ltd, Ardonagh, Walshestown, Mullingar, Co. Westmeath.
Leinster Chicks Ltd, Newcastle, Greystones, Co. Wicklow.
Tetra Chicks Ltd, Mount Avenue, Dundalk, Co. Louth.
Whitakers Hatcheries Ltd, Camden Quay, Cork.

South Africa

Pylon Products (PTY) Ltd, 12 Ramsay Street, Booyrens 2091, Johannesburg.
Stein Bros Ltd, 13 Springbok Way, P.O. Box 43157, Industria TVL 2042, Transvaal.
Tarton Poultry Farm (PTY) Ltd, P.O. Box 823, Randfontein 1760.

United Kingdom

Layers of White-Shelled Eggs
Babcock Farms Ltd, Goldlay House, Parkway, Chelmsford, Essex (BABCOCK B300).
Double-A Farms Ltd, Bentham Lane, Witcombe, Gloucestershire (HYLINE W36).
Euribrid Ltd, Chester High Road, Neston, Wirral, Cheshire/Distributors: Anglian Food Ltd (HISEX WHITE).
H and N Inc., 8 Buccleugh Street, Dalkeith, Midlothian (H & N NICKCHICK).
Ross Poultry Ltd, Poultry Stirling Division, Ross House, Grimsby, South Humberside (ROSS WHITE).
Shaver Poultry Breeding Farms (GB) Ltd, Bawdeswell, Dereham, Norfolk (SHAVER STARCROSS 288)

Layers of Brown Shelled Eggs
Anderson Poultry Breeding Organisation, Osborne House, Hawardsgate, Welwyn Garden City (ANDERSON ROCKET, ANDERSON BLONDE).
Arbor Acres (UK) Ltd, East Hanningfield, Chelmsford, Essex (ARBOR ACRES S.L.)

Babcock Farms Ltd, Goldlay House, Parkway, Chelmsford, Essex (BABCOCK B380).
Dekalb (U.K.) Ltd, Refuge House, 3 Kings Court, York (DEKALB AMBER LINK).
Euribrid Ltd, Chester High Road, Neston, Wirrall, Cheshire (HISEX BROWN).
Hubbard Poultry U.K. Ltd & Ireland, 54-56 London Road, Stroud, Gloucestershire (HUBBARD GOLDEN COMET).
Ross Poultry Ltd, Poultry Stirling Division, Ross House, Grimsby, South Humberside (ROSS BROWN).
Shaver Poultry Breeding Farms (GB) Ltd, Bawdeswell, Dereham, Norfolk (SHAVER STARCROSS 585).
Thornber Chicks Ltd, Mytholmroyd, Bank Lane, Hebden Bridge, Yorkshire (THORNBER).
Warren Studler Breeding Farm Ltd, Old Hall Hatchery, Orton Longueville, Peterborough, Cambridgeshire (WARREN STUDLER).

Broilers
Anderson Poultry Breeding Organisation, Osborne House, Hawardsgate, Welwyn Garden City (ANDERSON).
Anglian Poultry, High Street, Stalham, Norwich, Norfolk (COBB, GOLD, COBB 3).
Euribrid Ltd, Chester High Road, Neston, Wirral, Cheshire (HYBRO).
H and N Inc., 8 Buccleugh Street, Dalkeith, Midlothian (H & N CHICK).
Knayton Farms, Knayton, Thirsk, Yorkshire (WROLSTAD).
Bernard Matthews, Great Witchingham Hall, Norwich, Norfolk (MATTHEWS TEN 30).
Ross Poultry Ltd, Poultry Stirling Division, Ross House, Grimsby, South Humberside (ROSS 1, ROSS).
Shaver Poultry Breeding Farms (GB) Ltd, Bawdeswell, Dereham, Norfolk (STARBRO).

United States of America

Arbor Acres Farm Inc., Glastonbury, Connecticut 06033.
Dekalb Ag. Research Inc., Sycamore Road, Dekalb, Illinois.
Hy-Line Indian River Co., P.O. Box 6, Johnston, Iowa 50131.
Shaver Poultry Breeding Farms Inc, P.O. Box 759, Cullman, Alabama 35055.

Canada

Hybrid Turkeys Ltd, 9 Centennial Drive, Kitchener, Ontario.

United Kingdom

Attleborough Poultry Frams, Attleborough, Norfolk.
Arnewood Turkeys International Ltd, Banquet Kings Farm Ltd, Sway, Lymington, Hampshire.
A.M.S. Turkeys Ltd, Sandlow Green Farm, Holmes Chapel, Cheshire.
D.N. Barker Ltd, Walnut Tree Farm, Luffenhall, Near Stevenage, Hertfordshire.
Barron Turkeys Ltd, Thatched House Farm, Dutton, Warrington, Cheshire.
British United Turkeys Ltd, Stops House, 25 Nicholas Street, Chester, Cheshire.
Dales Turkeys Ltd, Affcot Hatchery, Marshbrook, Church Stretton, Shropshire.
Hy-Line Turkeys, Milestone Farm, Hereford Road, Ludlow, Shropshire.
Kelly's Turkeys, Springate Farm, Bicknacre Road, Danbury, Essex.
Leacroft Turkeys Ltd, Friars Close Farm, Barnwell, Oundle, Northamptonshire.
Spillers & Sons Turkeys Ltd, Manor Farm, Keyston, Huntingdon, Cambridgeshire.
Turners Turkeys Ltd, Spalding Drove, Claylake, Spalding, Lincolnshire.
Twydale Turkeys Ltd, Wansford Road, Driffield, East Yorkshire.
Edward Webster, Wash Farm, Ormskirk, Lancashire.
Wishbone Turkeys, Middleton Hall, Sudbury, Suffolk.

Index

Numbers in *italics* refer to black and white illustrations. Numbers in **bold** refer to colour plates.

174

175